T0140671

HIV-1 Proteomics

David R.M. Graham • David E. Ott
Editors

HIV-1 Proteomics

From Discovery to Clinical Application

Editors
David R.M. Graham
Department of Molecular and Comparative Pathobiology
The Johns Hopkins University School of Medicine
Baltimore, MD, USA

David E. Ott
Leidos Biomedical Research, Inc.
Frederick National Laboratory for Cancer Research
Frederick, MD, USA

ISBN 978-1-4939-8231-8 ISBN 978-1-4939-6542-7 (eBook)
DOI 10.1007/978-1-4939-6542-7

Printed on acid-free paper

This Springer imprint is published by Springer Nature
The registered company is Springer Science+Business Media LLC New York

We dedicate this book to Louis Edwin Henderson Ph.D. (1934–2013). Lou: a pioneer in retroviral protein chemistry, a mentor to so many, trusted colleague, a friend, and a tireless AIDS fighter.

Preface

The focus of this book is to provide a clear understanding of how the proteins in HIV-1 can be analyzed by proteomic approaches and cover some of the accomplishments produced by this powerful technique. Being in the field, we believe that much can be learned from the comprehensive study of proteins and their modifications. Throughout our combined careers, we have seen many scientific advances that were initiated by protein-focused research of HIV-1. To convey this appreciation for this topic, we have set out to compile expert views of HIV-1 proteomics. The target audiences for this book are those non-expert readers who desire a comprehensive but technically understandable view of the field, those in associated fields who want a deeper theoretical understanding and an entree into in-depth technical literature and resources through review references, and experts in the field who are interested in key points of interest and conflict in HIV-1 proteomics as well as current directions in the field. We hope you enjoy this work.

Baltimore, MD, USA — David R.M. Graham

Frederick, MD, USA — David E. Ott

Contents

Contributors

David R. Colquhoun, Ph.D. Molecular and Comparative Pathobiology, The Johns Hopkins University School of Medicine, Baltimore, MD, USA

David R.M. Graham, M.Sc., Ph.D. Department of Molecular and Comparative Pathobiology, The Johns Hopkins School of Medicine, Baltimore, MD, USA

Rebekah L. Gundry, Ph.D. Department of Biochemistry, Medical College of Wisconsin, Milwaukee, WI, USA

Shelby L. O'Connor, Ph.D. Department of Pathology and Laboratory Medicine, University of Wisconsin-Madison, Madison, WI, USA

David E. Ott, Ph.D. Leidos Biomedical Research, Inc., Frederick National Laboratory for Cancer Research, Frederick, MD, USA

Part 1
Rebranding Classical Protein Chemistry: Proteomics of the Past

Chapter 1
Introduction: HIV-1 Proteomics, Why Should One Care?

David E. Ott

This book focuses on applying proteomics to the study of the HIV-1 virion and its associated proteins. Before we go further, it is important to understand why this topic is worthy of a book or a researcher's effort to read it. Proteomics, the detection of the constellation of proteins and their posttranslational modifications in complex samples from organisms, has been used for years to comprehensively analyze protein samples to better understand basic biology and for medical research aimed at developing new and improved diagnostic procedures and therapies.

Despite the analytical power and potential of this important technology, these approaches have only recently been applied to the study of HIV-1 virions. HIV-1, like most RNA viruses, has a small genome, encoding just nine proteins. To compensate for this constraint, retroviruses use several unusual mechanisms to produce multifunctional proteins by differential splicing, alternative translational initiation/internal ribosomal entry site (IRES) usage, protease processing, translation frameshifting/nonsense codon suppression, and posttranslational modification [1, 2]. Many of these processes are well known now, but initially many of these now routinely understood cellular mechanisms for maximizing protein expression from a limited genome and modification in mammalian systems such as frameshifting [3] and myristylation [4] were initially observed in studying retroviral proteins. Thus, the study of retroviral proteins and their modifications has yielded many important discoveries. Yet these advances were made with traditional methods which, though still quite useful, remain constrained by relatively low sensitivity and limited throughput. Thus, even though much progress has been made in HIV-1 "proteomics," there are still many potential modifications and secrets of HIV-1 proteins

D.E. Ott, Ph.D. (✉)
Leidos Biomedical Research, Inc., Frederick National Laboratory for Cancer Research, 1050 Boyles Street, Frederick, MD 21702, USA
e-mail: ottde@mail.nih.gov

D.R.M. Graham, D.E. Ott (eds.), *HIV-1 Proteomics*,
DOI 10.1007/978-1-4939-6542-7_1

that remain to be discovered. High sensitivity and comprehensive mass spectrometry methods have the power and the capacity to search out and find these potentially undiscovered species and increase our understanding of HIV-1 and, in a wider sense, biology as a whole.

In addition to just the cellular proteins, HIV-1 also interacts with and sometimes acquires cellular proteins during assembly. These proteins can provide essential functions that support viral replication, either in assembly, infection, or immune evasion. Therefore, in addition simply the viral proteins in virus particles, the repertoire of the cellular proteins in the virion also can provide important information for understanding HIV-1 biology. Finally, HIV-1 as a lentivirus is a complex retrovirus with several proteins that modulate the various systems. Upon HIV-1 infection, expression of several of its proteins, some of which are not significantly incorporated into the virion, interferes the cellular signaling and cycling pathways. This represents a topic ripe for study with of various infected cell types with proteomic approaches to better understand the as yet elusive mechanisms for HIV-1 pathogenesis and cell killing.

In light of these aspects, proteomics offers a unique approach to:

- Understand HIV-1 viral assembly
- Understand HIV-1 infection/immune escape
- Understand HIV-1 pathogenesis
- Produce better drugs or therapies against HIV-1 and its related pathogenesis
- Design/produce effective vaccines

In light of these potential benefits, HIV-1 proteomics is a field that holds great promise to solve the various riddles and puzzles of this very important human pathogen and to combat the AIDS pandemic. Given all of these factors, the understanding how proteomics has been used and continues to be a vital tool for HIV-1 biology is important for anyone who wants a well-rounded view of HIV-1 research and biology. To this end, this book covers the proteomic analysis of HIV-1, its achievements, its challenges, its applications, and its potential for future directions.

References

1. Sundquist WI, Krausslich HG. HIV-1 assembly, budding, and maturation. Cold Spring Harb Perspect Med. 2012;2(7):a006924.
2. Karn J, Stoltzfus CM. Transcriptional and posttranscriptional regulation of HIV-1 gene expression. Cold Spring Harb Perspect Med. 2012;2(2):a006916.
3. Jacks T, Power MD, Masiarz FR, Luciw PA, Barr PJ, Varmus HE. Characterization of ribosomal frameshifting in HIV-1 gag-pol expression. Nature. 1988;331(6153):280–3.
4. Henderson LE, Krutzsch HC, Oroszlan S. Myristyl amino-terminal acylation of murine retrovirus proteins: an unusual post-translational proteins modification. Proc Natl Acad Sci U S A. 1983;80(2):339–43.

Chapter 2
HIV-1 Biology at the Protein Level

David E. Ott

The Benefits of HIV-1 Protein Study

To understand a biological entity, one primarily needs to understand its nature and composition. Thus, one of the first experiments carried out after identifying a new virus is to examine its protein content because that is the critical part of the virus that will lead to an understanding of its biology, open up ways for diagnosis, and guide the development of antiviral strategies. Yes, we rightly classify viruses based on their genome: DNA, RNA, double strand, single strand, plus, or minus, and these nucleic acids encode proteins as well as other genetic elements such as microRNAs which, along with proteins, regulate gene expression. Yet, it is the proteins in the virions that catalytically carry out the replication cycle, induce immune responses that can be helpful or detrimental to the host, cause pathogenic effects in the host, and provide for vaccine and drug therapeutic targets. One of the reasons why HIV-1 antivirals were produced so rapidly after the discovery of HIV/AIDS was that we already had an extensive understanding of retroviruses from data gathered from intensive protein analysis of avian, murine, equine, feline, and other viruses, which provided the basic understanding of retroviral biology, from how they replicate and their essential enzymatic reactions to the serological responses against them that are induced in the host. It is important to note that retroviruses led to the discovery of oncogenes since they can recombine with and, thus, transduce cells with cellular regulatory genes that were acquired by the virus during reverse transcription [1, 2]. In fact, before the term retroviruses was coined, most were isolated from tumors or leukemias, e.g., Rous sarcoma virus and murine leukemia virus, and placed under the umbrella term RNA tumor viruses. Thus, retroviral protein studies also greatly

D.E. Ott, Ph.D. (✉)
Leidos Biomedical Research, Inc., Frederick National Laboratory for Cancer Research, 1050 Boyles Street, Frederick, MD 21702, USA
e-mail: ottde@mail.nih.gov

D.R.M. Graham, D.E. Ott (eds.), *HIV-1 Proteomics*,
DOI 10.1007/978-1-4939-6542-7_2

contributed to the genetics of cancer, signal transduction, differentiation, and cell cycle control [1, 2]. The previous groundwork with retroviruses also saved many lives during the early stages of the AIDS crisis by accelerating the design and production of the HIV-1-induced antibody screening test [3]. The rapid development of this test, from virus discovery in 1983 [4] to commercial test in 1985, reduced an immeasurably large number of new infections by protecting the donated blood supply and providing diagnostic screening for at-risk people to identify AIDS carriers. Additionally, biochemical understanding of the reverse transcriptase process resulted in the use of the first antiviral, AZT/zidovudine, in 1987 [5, 6] only 4 years post discovery, a remarkably rapid drug development for a new pathogen. Later, more effective antiviral therapies targeting protease [7] were approved in 1995 [8], and the early forms of the current clinical antiviral therapy, highly active antiviral therapy (HAART aka CART) [9], were developed soon after in 1996. HAART uses combinations of multiple classes of drugs that initially targeted reverse transcriptase and protease by different mechanisms to combat the mutation-driven HIV resistance to single-drug therapy. Viruses developing resistance to one class have to also have resistance to the others to replicate efficiently. Antivirals against other targets including integrase [10] and Env [11] have been developed and are now used in HAART/CART regimens. The foundation of all of these advances is based on an impressive understanding of the basic biochemistry and structure of the viral proteins.

Evolution of Retroviral Biochemistry

The biochemical study of retroviruses began during the late 1960s and evolved through the 1990s [12–16]. Please note that in this book, the term retroviruses will refer to only orthoretroviruses which include HIV-1. The spumaretroviruses/foamy viruses [17], which are similar yet have several important replication differences, are not discussed in this book. Initially, proteins were studied by rudimentary techniques. Indeed, the essential retroviral biochemistry study that uncovered the basis for the paradigm-shifting discovery of a polymerase, retroviral reverse transcriptase, that reverses the “central dogma” of molecular biology was achieved by classical biochemical enzymatic assays on extracts of partially purified murine and avian retroviral particles [18, 19]. The advent of SDS-polyacrylamide gel electrophoresis in 1965 revolutionized protein analysis, greatly improving separation and resolution by molecular mass [20]. Applying this to retroviral virions provided the first looks into its proteome. In fact, the retroviral proteins are still referred to by some by their molecular mass, e.g., p24 or gp120. Subsequent advancement in technology such as large-scale virus purification combined with immunoblotting, metabolic radiolabeling, Edman degradation sequencing, and amino acid analysis allowed for a remarkable characterization of the proteins in retroviral particles. Contrary to the ease at which nucleic acids are sequenced to deduce protein sequence today, many of the first full sequences of retroviral proteins were produced by painstaking protein sequencing methods. Several protein modifications of viral proteins by the cell as well as the viral protease processing sites were also uncovered by these protein sequencing analyses, information that cannot be revealed by sequencing DNA. Therefore, even though amazing technical nucleic acid efforts, such as the human genome project, have provided a wealth of information, studying the proteins themselves is still vital as the virus has many tricks that it plays with its genome. So although nucleic acid-based methods reigned in biology in the late 1980s through the mid-2000s, the analysis of the proteins, the active agents in the cell, in the form of high-power mass spectrometry to carry out "proteomic" analysis, i.e., a detailed and refined analysis of each individual protein in a complex mixture, ushered in new appreciation for protein study.

The analysis of HIV-1-associated proteins has been a crucial part of studying human immunodeficiency virus type 1 from the beginning. At the time of the discovery of HIV-1 in 1983, it was fortunate that the methods and ability to analyze retroviral proteins were already in place. These consisted of what would now be considered classical biochemistry: immunoblots, immunoprecipitation, column chromatography, amino acid analysis, and degradative protein sequencing (presented in this chapter). In contrast, mass spectrometry was mostly confined to relatively small molecules. Unlike today, the rather primitive computers of the day also played a role in limiting its technical abilities. Rapidly applying biochemistry steered by the prior knowledge of retrovirology to this "new" virus that caused AIDS brought forth a fountain of basic information about HIV-1: protein makeup of the virion, protein sequence, identification and characterization of critical of enzymes, and detection of new retroviral proteins. In turn, this information along with contributions from other fields generated the AIDS test and drugs that have saved so many lives. While classical biochemistry remains a powerful set of techniques that are still invaluable to protein analysis, mass spectrometry techniques have breathtakingly evolved (presented in Part II), emerging as a dynamic, multifaceted tool that allows for highly sensitive, high-throughput analyses of proteins present in complex mixtures, approaches often placed under the banner proteomics. Yet despite these new courses of study made possible by advanced mass spectrometry, the classical approaches either by themselves or in conjunction with mass spectrometry still remain vital tools. There are experiments that mass spectrometry still cannot carry out. Thus the classical and the new are more complementary rather than redundant.

HIV-1 Virus Genome and Its Proteins: Basics

Retroviral particles are composed of proteins, RNA, and lipids [21]. While RNA, both viral genomic RNA and a host primer tRNA, and lipids are required for infectious particle formation, the viral proteins in the virion do the critical work in replication and, thus, are the most studied components of the virion. As with most RNA viruses, genome coding capacity is at a premium due to the instability of RNA; therefore, HIV-1 produces these proteins in an efficient, temporally managed fashion that is a tour de force in economy and design, using nearly all of the tricks available in mammalian biology that provide for the parsimonious production of protein activities: protease processing, frameshifting, differential RNA splicing, regulated mRNA nuclear export, overlapping ORFs, alternative translation initiation, and internal ribosomal entry sites [22–24]. HIV-1 produces one unspliced RNA encoding multiple proteins in all three reading frames (Fig. 2.1) and six spliced RNAs. Using several approaches, these RNAs produce three polyproteins, each essential for infectivity, as well as six additional proteins that assist in HIV-1 replication. Thus, the virus-encoded proteome of HIV-1 appears to be relatively small, nine proteins (Fig. 2.1) that include polyproteins which are further processed

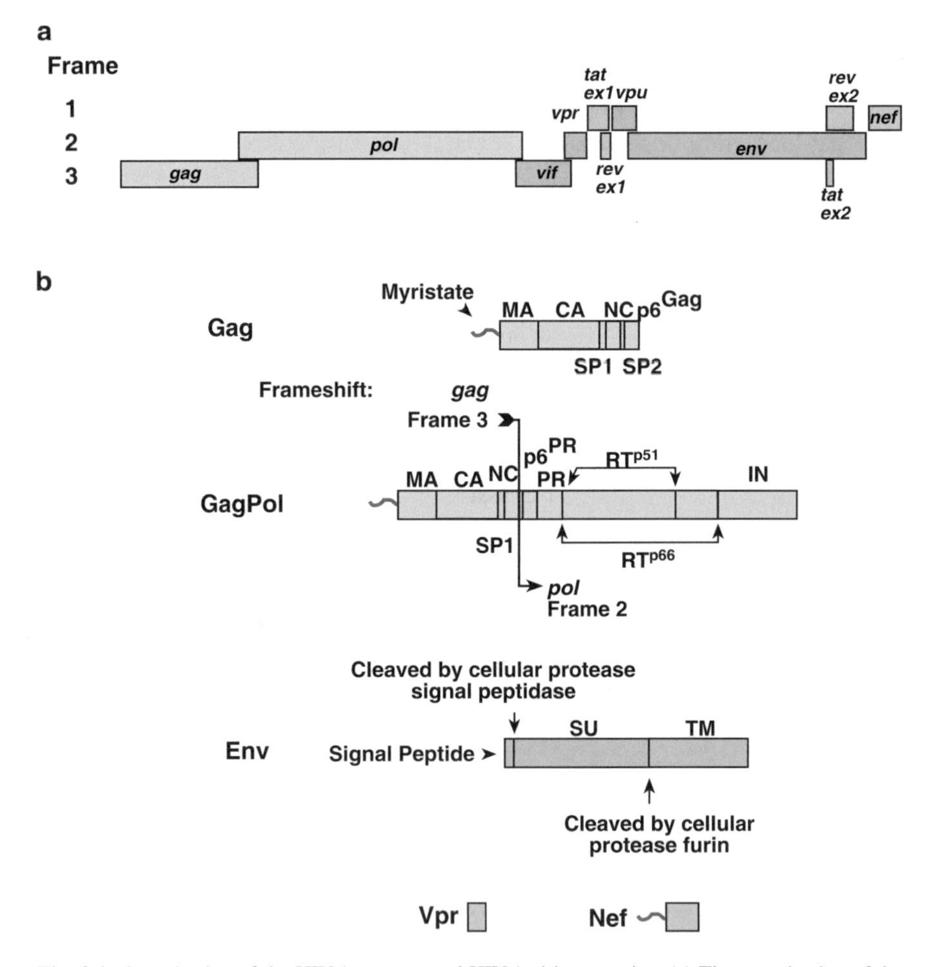

Fig. 2.1 Organization of the HIV-1 genome and HIV-1 virion proteins. (**a**) The organization of the HIV-1 genomic open reading frames is presented with proteins expressed from unspliced genomic RNA in *orange*, single-spliced RNA in *green*, and multiply spliced RNA in *blue*. (**b**) Diagram of the viral proteins incorporated into virions. Unless indicated otherwise, all internal *vertical lines* in protein diagrams denote HIV-1 protease cleavage sites. The GagPol frame shift site is denoted by an *arrow* and labeling. Color scheme for RNA splicing is as above. Myristyl modification is denoted in red

by proteases to provide for the full major complement of 22 distinct peptides from which 15 functional proteins are made [23, 24]. All of the essential structural proteins, i.e., those minimally required within an infectious particle, are produced first as polyproteins, which are cleaved into several smaller mature proteins by a coordinated process that ensures proper folding to carry out replication. These critical proteins, Gag, GagPol, and Env, are expressed, processed, and incorporated into virions by all retroviruses. In addition to the Gag, GagPol, and Env polyproteins and their mature processed forms, two proteins, Vpr and Nef, that are unique to HIV-1 and the closely related HIV-2 and simian immunodeficiency virus (SIV) are also incorporated into virions. The other four initially expressed proteins act in the infected cell, enhancing and regulating transcription/splicing [22] as well as altering cellular processes and defeating innate defenses [25, 26]. The remainder of this chapter will focus on those proteins that are found in the virion, thus readily studied by protein analysis of virus particles.

Viral Proteins in HIV-1 Virions

The most abundant protein in the virion is Gag and its mature protein products [15, 16, 27]. Gag is the only retroviral protein that is strictly required to produce "virus-like" particles; Gag drives particle formation through strong intermolecular interactions with other Gag molecules, with RNA, and with plasma membrane lipids through which the particle buds acquire a host-derived membrane envelope. During this budding process, the viral surface glycoprotein, Env, which provides for target cell binding and entry into the host cell, is also brought into the virion with the plasma membrane [15, 16, 27]. The Env precursor is incorporated into the virion at lower levels than Gag [28], cleaved in the late Golgi apparatus, forming a surface glycoprotein (SU) that is noncovalently attached to a transmembrane (TM) protein that trimerizes to form Env complexes. Env trimers on the surface of the virion bind receptors (CD4 and CXCR4 or CCR5) on host cells and induce a fusion event, which empties the infection machinery into the cell (Fig. 2.2).

Unlike the Gag proteins, only small amounts of the enzymatic proteins of the virus are needed [15, 16, 27], protease (which processes Gag and GagPol during virion maturation), reverse transcriptase (which converts the genomic RNA into viral DNA using a complex series of DNA polymerization steps), and integrase (which places the viral DNA into the host chromosomal DNA to form the provirus that expresses the viral RNAs for another round of replication). To express the small amounts of these Pol proteins required, about 5 % of the Gag translations undergo a -1 frameshift that redirects the ribosome from the *gag* gene reading frame to that of the *pol* gene (Fig. 2.1). This results in a GagPol polyprotein that can join the forming particles with Gag due to intermolecular interactions between Gag. Because both Gag and GagPol are expressed polyproteins that assemble into the virion, the correct ratio of the structural and enzymatic proteins is incorporated into the particle. Both of these polyproteins are subsequently processed by

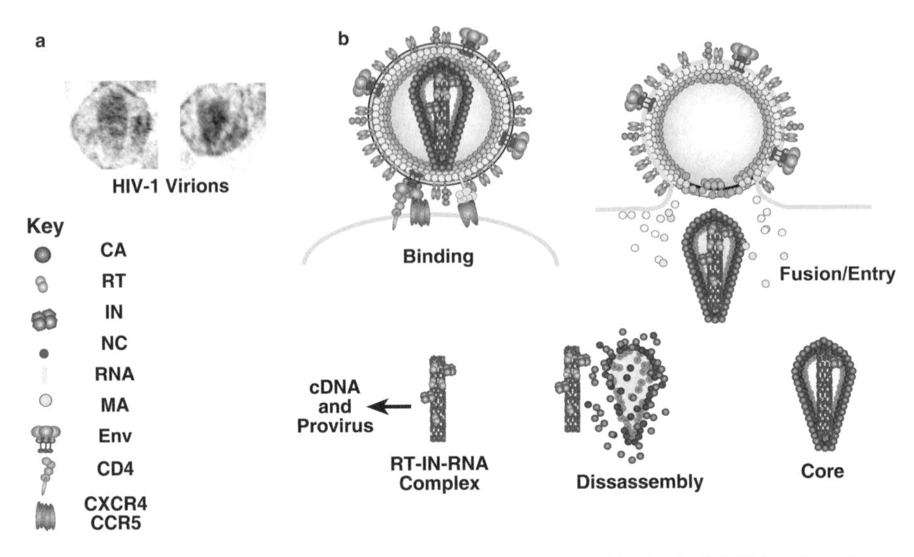

Fig. 2.2 HIV-1 entry and uncoating. (**a**) Electron micrographs of individual HIV-1 virions showing the conical core structure are presented. (**b**) A diagram of HIV-1 host binding through uncoating is presented. A color key identifies the virion proteins, and the HIV-1 receptor, CD4, and coreceptor, CXCR4/CCR5 are presented at left

protease during and after assembly [15, 16, 27]. Processing is a sequential reaction liberating the mature proteins in a coordinated fashion that results in a structural reorganization of the virion interior, forming a mature, fully infectious virion with a conical core [29].

In addition to the essential Gag, GagPol, and Env proteins, the other virally encoded proteins incorporated into virus particles, Vpr and Nef, are referred to, as accessory proteins. Despite the accessory name, these proteins, Vpr, Vif, Vpu, and Nef, are nonetheless essential for robust in vivo infection and pathogenesis [30]. They are only "nonessential" when examined in artificial cell culture-based assays that do not accurately reflect in vivo conditions. The strong selective pressure to economize the relatively low coding capacity genome of RNA viruses, especially retroviruses, ensures that all of the viral proteins are "essential" for the virus in vivo; otherwise, they would not have been maintained in the viral genome. Thus, the nonessential label arises from the artificial perspective of in vitro reductionist studies and in comparison with simpler retroviruses that express the three universal retroviral proteins, such as the prototypic avian retroviruses. Of the four proteins, Vpr, which is thought to assist nuclear entry of the PIC during infection (see below), is incorporated into virions at the greatest extent, ~ 14 % of Gag [31], as fitting for a protein that functions in the target cell. Vpr is brought into the virion through an interaction with the C-terminal region of the Gag polyprotein, sequences in $p6^{Gag}$ [32]. In contrast, Nef, which is mostly believed to act at the productively infected stage of replication, is also incorporated into the virion but in relatively small amounts, consistent with its role in virus production. Nef seems to be incorporated due to its strong binding to the plasma membrane, being a passive bystander caught up in the budding process. The necessity of incorporating the other proteins into virions is less clear. From biochemical analyses, Tat, Rev, Vif, and Vpu are typically thought to be excluded for the virus particle. Moreover, any reported HIV-1 packaging of these proteins did not appear to be the result of a specific incorporation [33] or have any functional consequences.

HIV-1 Assembly: Why Gag Polyproteins?

Particle assembly relies on getting the right proteins in the right place with the right structure. Due to the large size of the cell and the complexity of the cytoplasm, proteins that are destined to assemble into the virus particle need to have both temporal-spatial and structural coordination. HIV-1 has eight proteins that are required for assembly of an infectious virion: MA, CA, NC, $p6^{Gag}$, PR, RT, IN, and Env, SU, and TM (these proteins are discussed below). If these many parts are expressed individually, then virion assembly would face daunting problems of specific transport, coordinated expression, and proper folding and oligomerization in the chaotic environment of the cell. Combining expression of these proteins into units that blend their properties together in a polyprotein overcomes these issues. Proteins with related roles in replication are linked together for transport so only one protein needs to find its way to the plasma membrane. Polyprotein expression ensures that these proteins are present at the proper levels and in an equal stoichiometry, something especially important for building complex structures. Finally, putting like proteins together allows them to fold coordinately, each making its own intra-subunit interaction with inter-subunit interactions being made easier by cooperative binding between all the subunits in the polyprotein. The polyprotein is an efficient strategy to solve the temporal-spatial and structural demands of assembly.

Thus, HIV-1 Gag as the primary driver of assembly contains all of the proteins, MA, CA, and NC, that provide for the structure of the virion [15, 16, 27]. These subunits, as domains in the Gag polyprotein, interact with the same subunits in other Gag molecules or cellular membranes and RNAs to make it fold into a rodlike structure that possesses both strong plasma membrane binding and oligomerization properties [29]. Upon binding of Gag to the plasma membrane, the strong multimerization properties of Gag organize into larger complexes of Gag hexamers that structurally tie themselves to neighboring hexamers to produce sheets that begin to curve, ultimately producing a spherical particle that buds from the plasma membrane [29].

The virion must have its enzymes for replication, protease, reverse transcriptase, and integrase, but the amount needed is catalytic not structural. In fact, overexpression of GagPol results in premature processing before assembly [34]. Thus, the Pol region is expressed at a much lower level (5 %) than full-length Gag [15, 16, 27]. The strategy of expressing Pol as a frameshift of Gag achieves two important functions; Pol is produced at an appropriately low level and the N-terminus of Pol is fused to Gag allowing for its incorporation into the particle. Also, because Gag and GagPol are produced for the same polysomes, they can interact and assemble together at the same place in the same time in the right amounts, solving the temporal-spatial and stoichiometric problems. Overall the Gag and GagPol polyproteins are an ingenious strategy used by retroviruses to provide for the incorporation of the essential mature proteins at the correct ratios in the correct structures, rather than doing these functions for each mature protein independently.

Gag and GagPol are mostly assembly machines, possessing strong intermolecular interactions that are very stable [15, 16, 27]. However, infectivity requires the virion proteins to disassociate in order to carry out reverse transcription and integration. So this assembly machine needs to switch to an infection machine. To do this both Gag and GagPol subunits are processed into mature proteins by the Pol-encoded protease both during and after virus release from the cell [35]. Protease processing unleashes the poised subunit structures that, when together in the polyprotein make Gag strongly multimeric, once liberated from each other by protease go their own way forming homotypic interactions and structures that completely restructure the virion and transform it into an infection machine. The mature structure is a particle with a HIV-1 trademark conical capsid core inside the virion [29] (Fig. 2.1). This maturation process is not random as processing occurs in a coordinated fashion that regulates the formation of the different structures, performing a molecular ballet with everything occuring at its proper time. Coordination of processing is critical as alteration of the normal order of stepwise processing results in a noninfectious particle [35]. From assembly through infection, the polyprotein strategy is a sophisticated, elegant, and even beautiful dynamic process used by retroviruses to solve the logistical problems of many proteins, a small genome, and a big cell.

As a polyprotein, Env transverses the secretory system as any transmembrane protein does where it is processed in the Golgi apparatus into the Env molecule (see below). The expression of Env as a polyprotein is most likely due to the need for both subunits to assemble together. The mechanism for Env incorporation into the virion is currently not clear, though it seems to be passively incorporated [15, 16, 27].

The HIV-1 Proteins

Below are brief summaries of the proteins produced by HIV-1 with an emphasis on those incorporated into the virion. Accompanying each entry is the web address for its UNIPROT database reference which contains valuable information on function, modifications, processing, sequences (protein and data), and helpful references.

Gag (http://www.uniprot.org/uniprot/P12493)

The Gag polyprotein is produced from the unspliced full-length RNA species that serves as both an mRNA and the viral genome (reviewed in [36]). As discussed above, Gag produces all of the proteins sufficient for the production of particles and drives particle formation, though expression of Gag only produces particles, viruslike particles or VLPs, that are not infectious [15]. The bulk of the protein within virions is Gag, each virion containing approximately 2500 copies of Gag protein [37], consistent with its role as the major structural component of the virion. Gag is cleaved by the GagPol-encoded protease, reviewed below, into six mature proteins, MA ($p15^{MA}$), CA ($p24^{CA}$), SP1 (also known as p2), NC ($p7^{NC}$), SP2 (also known as p1), and $p6^{Gag}$, in a process initiated at the start of viral assembly and completed after particle release from the cell (Fig. 2.1). The Gag proteins fall into three categories: (1) those that have a function in the mature virion and infection, MA, CA, and NC; (2) those that act in Gag only, regulating the coordination of Gag processing, SP1and SP2; and (3) $p6^{Gag}$ which interacts with cellular proteins to assist virion release from the cell [15, 16, 27]. Stepwise processing of both the Gag polyprotein and the GagPol protein (discussed below) is strictly required for infectivity of the virus particles release from the cell [35]. In addition to these major proteins, there are also minor proteins that are produced from low levels of internal cleavage of several mature Gag proteins, forming fragments of SP1, SP2, and $p6^{Gag}$ [38]. Their importance in viral biology is currently unclear. As presented above, the individual mature Gag proteins play different, yet similar roles as both subunits in Gag and protein in the mature virion. Below is a brief description of the mature Gag proteins (note: UNIPROT does not contain individual entries for the mature Gag proteins. Please use the Gag entry for them).

MA, Matrix, $p17^{MA}$

This N-terminal protein in Gag is N-terminally myristylated cotranslationally during Gag synthesis. The addition of myristate to the MA subunit in Gag is essential for Gag targeting to the plasma membrane and particle formation [15, 16, 27]. Matrix functions primarily at the assembly stage, directing the localization and membrane binding of the Gag polyprotein by the N-terminal myristyl fatty acid modification, a positive stretch of amino acids, and a phosphatidylinositol-4,5-bisphosphate [PI(4,5) P_2] binding site [39]. These features coordinately cause Gag to be strongly attached to the membrane with a hexameric arrangement of MA trimers [29]. The interactions between the different regions of Gag are complex and beyond the scope of this review. Informative reviews that detail these Gag-Gag assembly interactions and the transition of Gag in particles from its "immature" polyprotein state to its mature infections structure are available [27, 29, 40]. MA in mature virions is loosely associated with the virion membrane. It is not part of the core structure that contains the genome and the essential enzymes for reverse transcription and integration. In addition to myristylation, MA is phosphorylated near a nuclear localization site. Though this was once thought to be important for infection in nondividing cells, phosphorylation of MA does not seem to play an obvious role in replication [41]. The structural organization of the membrane-associated MA hexamers, presumably in the context of Gag, appears to limit the incorporation of Env trimers into the virion by steric hindrance: the long cytoplasmic tails of Env can only fit through the holes in the hexameric lattice made up of MA trimers, rather than a direct and positive MA-Env interaction that drives Env incorporation [42, 43].

CA, Capsid, or p24^{CA}

CA as a region in Gag provides a strong protein-protein interaction force that is critical to particle assembly. Similar to MA, the N-terminus of the CA subunit interacts to produce hexamers, while the C-terminus forms dimers, apparently both within and between Gag hexamer arrays knitting the Gag proteins together [29]. During the proteolytic processing of Gag, sequentially liberated forms of CA rearrange stepwise [35] to gradually produce a cone-shaped core that consists of ~1500 CA proteins, mostly formed as an array of about 250 hexamers with 12 pentamer contacts [29, 44]. This conical structure within HIV-1 particles contains both the dimeric RNA genome and the enzymes required for establishing an infection/provirus, reverse transcriptase, and integrase [45, 46]. Upon fusion of the viral and cellular membranes, the CA cores enter into the cell and later uncoat and disassemble some of the capsid structure, allowing reverse transcription to begin (Fig. 2.2) [46]. Proper formation of the core is critical for the production of an infectious virus [47]. The requirements for CA-CA interactions are relatively strict such that many CA mutations drastically alter the formation of the core. The impact of CA is such that small molecules that interrupt CA processing are being evaluated as anti-HIV-1 therapies [48]. Some cells appear to have an innate immunity that targets CA. There are several cellular proteins, APOBEC3G, Tetherin, and TRIM5α, collectively termed restriction factors, which inhibit HIV-1 replication [49, 50]. Of these, TRIM5α from other species, e.g., rhesus macaque, restricts HIV-1 infection by binding to CA cores. The exact mechanism for this restriction is not known, but it appears that TRIM5α binding advances the timing of core uncoating either directly or through degradation of CA by the ubiquitin proteasome pathway [51, 52], interfering with reverse transcription and integration. CA interacts with cyclophilin A in the infecting cells, and interrupting this interaction can reduce HIV-1 infectivity, but the mechanism is unclear and complicated [53]. CA is phosphorylated on several serines. Although some mutagenic data suggest somewhat that some of these sites are important in CA function, the general intolerance of CA to substitutions makes this finding only suggestive [54, 55]. Thus, the importance of this posttranslational modification is not clear.

SP1, Spacer Protein 1, p2^{Gag}

SP1, which was previously known as p2 due to its apparent molecular mass, appears to play a role in both the assembly of the spherical immature particle and the proper formation of the mature conical core. Mutations in the first four amino acids in SP1 within Gag drastically alter the formation of normal particles in the absence of protease, implying that SP1 acts as a molecular switch to provide an organizing function to immature virion assembly [56]. On the maturation side, the CA-SP1 protease cleavage site is the last to be cleaved by the HIV-1 protease in the normal sequence of processing [35], releasing SP1 from the C-terminus of CA. Removing SP1 from CA allows it to complete the final stages of core formation [47]; thus this processing controls the timing of the CA-CA interactions that allows for the appropriate conical core formation. Mutations that interfere with the timing of cleavage, either accelerating or preventing cleavage, and small-molecule compounds that block processing at this site form misshapen cores due to the retention of the CA-SP1 structure, yielding virions that are noninfectious [35]. Thus, the primary function of SP1 is to coordinate and regulate the intricate transition of CA from a region that knits Gag together in the spherical immature particle to one that produces the highly ordered CA shell that produces the conical core. SP1 and its cleavage are the target of the core interrupting compounds [48].

Nucleocapsid, NC, $p7^{NC}$

NC is a highly basic protein that binds both RNA and DNA and which plays several different roles in HIV-1 replication [57, 58]. NC as a portion of Gag assists in assembly by binding RNA to provide a scaffold that constrains Gag to bring the monomers into close contact to promote the CA-CA and MA-membrane interactions that drive assembly [59]. The NC region of Gag also selectively binds unspliced HIV-1 RNA to package it as dimeric genomic RNA into the assembling virion [15, 16, 27, 57]. NC is present in the viral core, binding the genomic RNAs. NC also plays a role in reverse transcription, assisting the formation of the viral cDNA from the genomic RNA by acting as an RNA chaperone that alters the genomic RNAs. By melting RNA secondary structure and enhancing formation of alternate structures, NC provides valuable assistance to the complex synthetic molecular gymnastics that occur during the reverse transcription of the genomic RNA into a full-length viral cDNA [46, 60]. Even after cDNA formation, NC still plays an as yet undefined role in integration [58], the formation of a provirus in the host chromosomal DNA (see Integrase section below). NC has not been found to have significant posttranslational modifications.

$p6^{Gag}$

As the C-terminal portion of Gag, $p6^{Gag}$ plays two main roles in assembly. (1) $p6^{Gag}$ binds several cellular proteins, notably Tsg101 and ALIX, that, in turn, stimulate the release of the budding particle from the cell surface through the use of the cellular budding machinery [27, 61]. The $p6^{Gag}$ protein does not seem to have any post-assembly function and is not found in the capsid core [62]. Additionally, the $p6^{Gag}$ in the Gag polyprotein binds the viral Vpr protein causing it to be packaged into the assembling virion [32, 63]. A fraction of the mature $p6^{Gag}$ is ubiquitinated, sumoylated, or phosphorylated. The role of the small amount of ubiquitin attached to Gag (~2 %) in the virus release remains controversial [64, 65] with some studies favoring an active role of ubiquitinated Gag in release while others support ubiquitination as a by-product, being more of a consequence of an interaction with a ubiquitinating activity as it uses the cellular release machinery [65]. Similar to ubiquitin a small amount of Gag is covalently modified with Sumo-1, though this seems to have a potentially negative effect in infection rather than assembly [66]. The $p6^{Gag}$ protein is phosphorylated on several serine and threonine amino acids; however, the importance of this modification is currently unclear [67, 68].

GagPol (http://www.uniprot.org/uniprot/P12497)

The GagPol polyprotein is produced from the same unspliced RNA as Gag by a -1 ribosomal frameshift that occurs in ~5 % of the Gag translations just after NC (reviewed in [36]). This results in a shift from the *gag* to the *pol* open reading frame, replacing the *gag*-encoded SP2 and $p6^{Gag}$ proteins with the *pol*-encoded $p6^{Pol}$ (also known as the transframe protein or preprotease), protease (PR), reverse transcriptase (RT), and integrase (IN) proteins (Fig. 2.1). GagPol is brought into the assembling virion by Gag interacting with Gag portion of GagPol. There are approximately 125 copies of GagPol proteins in the mature virion, consistent with its role providing the enzymatic component of the virion. Similar to Gag, GagPol is cleaved by the HIV-1 protease, initially with protease as a subunit with low activity that, upon cleavage, liberates highly active mature PR that then finishes GagPol processing concurrently with Gag to produce eight mature proteins, MA, CA, SP1, NC, $p6^{Pol}$, PR, RT, and IN (Fig. 2.1) [35]. Similar to Gag, the Pol portion of GagPol also undergoes sequential processing [69]. An overview of the proteins encoded by the Pol ORF is presented below.

$p6^{Pol}$, Preprotease, Transframe Protein

The $p6^{Pol}$ protein appears to function simply as a regulatory sequence for PR since mutations in this region alter the initial processing of PR from the GagPol polyprotein [70].

PR, Protease

PR is the viral aspartic protease that, acting as a dimer, processes both the Gag and GagPol polyproteins [15, 16, 27]. The sequence requirements for the PR recognition site are fairly promiscuous [71] with optimal sites generally being cleaved before those with less optimal characteristics. This property, along with steric hindrance, causes PR to process the polyproteins in an ordered fashion which, in turn, induces the proper folding of the mature Gag and Pol proteins into the functional virion as present above [35]. PR processing might also be regulated by reversible oxidation (inactive) and reduction (active) of cysteines in PR [72]. PR is a main target for anti-HIV-1 therapies such as HAART/CART.

RT, Reverse Transcriptase

The RT enzyme provides the RNA-dependant DNA polymerase that converts the dimeric HIV-1 RNA genome into the substrate for integration: a linear cDNA structure with two long terminal repeats (LTRs) at each end [46]. RT is a heterodimer of a p66 subunit and a p51 subunit, the latter produced by a differential cleavage of p66 (Fig. 2.1). The reverse transcription of the genomic RNA carried out by RT involves a complex process that generates a full-length cDNA with long terminal repeats, LTRs, on each end from the mRNA-like structure of the genomic RNA [46]. RT is found associated with NC and the genomic RNAs in the virion core structure (Fig. 2.2) [27]. After reverse transcription, the remnants of the core organize into the preintegration complex, also referred to as a PIC which contains the viral cDNA genome, IN, and Vpr, that carries out the final infection step, integration of the provirus into the host genome [73].

IN, Integrase

The IN protein is found in the core as a tetramer (Fig. 2.2). The IN binds to the LTR ends and integrates the viral cDNA into the host genomic DNA through a mechanism of staggered cuts, ligating one strand of the cDNA to the chromosomal DNA and relying on cellular DNA repair to ligate the second strand [45]. Integration also relies on cellular proteins to allow the preintegration complex to bind to and access the chromatin structure to produce the provirus stage of the replication cycle [45, 73]. In nondividing cells, the IN within the preintegration complex can interact with several cellular proteins allowing it to transverse the nuclear pore and enter the nucleus to access the host chromosomal DNA [45, 73].

Env (http://www.uniprot.org/uniprot/P03377)

The Envelope protein (Env) is the surface glycoprotein protein complex which resides on the surface of HIV-1 virions that allows the core to enter the host cell (Fig. 2.2) and begin the infection process [42, 74]. The Env glycoprotein complex consists of a trimer of $gp120^{SU}$ (SU)/$gp41^{TM}$ (TM) dimers (Fig. 2.2). Env is translated as a gp160 complex that is processed in the Golgi compartment by furin or a furin-like protease

(Fig. 2.2) [42]. The trimeric SU/TM complex, referred to as Env, binds CD4 on HIV-1's target cells which induces a conformational change that opens up a coreceptor-binding site in SU that, in turn, allows Env to bind a host surface protein, a chemokine receptor CXCR4 or CCR5 [42, 74]. Coreceptor binding exposes the fusogenic regions of the TM protein that fuses the virion membrane with that of the target cell, spilling out the contents of the virion allowing the HIV-1 core to enter and start the infection process [42, 74]. SU is heavily glycosylated, a feature that shields it from host antibodies [15, 74]. TM is also glycosylated, though less so than SU, and has two hydrophobic alpha helices arranged in helical bundles, one in the ectodomain, located in the extracellular space that interacts with SU, and the other in the membrane-spanning region [42, 74]. Upon coreceptor activation, the helices in the ectodomain penetrate the target cell plasma membrane. TM draws the two sets of helices together causing virion and target cell membrane fusion [42, 74, 75]. Slightly less than half of TM is a cytoplasmic tail which has several cellular protein binding functions [76] and is palmitoylated at two sites [77], though their significance is not clear.

Accessory Proteins

In addition to the standard Gag, Pol, and Env proteins present in all orthoretroviruses, HIV-1 has several proteins typically described as accessory proteins that, while not strictly required for replication in vitro, are nevertheless required for efficient transmission, replication, and pathogenesis in vivo [25]. Since the focus of this book is the proteomic analysis of HIV-1 virions, only the accessory proteins significantly incorporated into virions, Vpr and Nef, will be discussed at any length. However, Vif (http://www.uniprot.org/uniprot/P12504) and Vpu (http://www.uniprot.org/uniprot/P05923) play important roles in HIV-1 biology, especially in counteracting intrinsic host antiviral defenses, and should be studied by those wishing to understand HIV-1 biology [78].

Vpr (http://www.uniprot.org/uniprot/P12520)

The 11 kDa Vpr protein provides several important functions in HIV-1 replication [79]. About 250 copies of Vpr are packaged per virion through a specific interaction with p6^{Gag} [31] and Vpr is found in the core [62]. One of the most important functions is its ability to bind to and assist the import of the preintegration complex into the nucleus of nondividing and cell-cell cycle arrested cells. Integration of the viral cDNA into the host chromosomal DNA is required, yet is normally prevented by the nuclear membrane. Two ways in which the HIV-1 can access the host genome are (1) during cell division when the nuclear membrane is broken down and (2) through nuclear import of the preintegration complex by interactions of IN and Vpr with the nuclear pore [73, 80]. Vpr is required for the infection of cells that although active do not divide such as macrophages, which are important in vivo targets for HIV-1 infection and pathogenesis [81]. Vpr also arrests dividing cells at the G_2 stage of the cell cycle, induces apoptosis, and produces other pathogenic effects [82]. The functional implications of these properties are not clear, but likely increase replication, transmission, and possible immune evasion.

Nef (http://www.uniprot.org/uniprot/P03406)

The Nef protein is packaged in only low amounts [83], and most of its properties would not mechanistically require its incorporation: CD4, MHC I, and MHCII downregulation as well as altering cell signaling by binding PAK2 [84]. Nef is also myristylated and binds the plasma membrane so Nef incorporation may be simply due to passive packaging during budding. Nef expression in the producer cell increases the infectivity of the resultant virions, though whether this requires Nef incorporation into the virion remains controversial [85].

Regulatory Proteins

Finally, there are two transcriptional regulatory proteins expressed by HIV-1 that are critical for replication, Tat (http://www.uniprot.org/uniprot/P04610) which transactivates transcription from the HIV-1 promoter and Rev (http://www.uniprot.org/uniprot/P04618) which exports the unspliced, full-length mRNA genome and singly spliced mRNA from the nucleus into the cytoplasm to provide both genomic RNA and Gag and GagPol as well as Env and Vpu expression [86]. While the biology of these proteins is very interesting and important, neither of these appears to be incorporated into the virion, thus beyond the focus of this book. Please see these excellent reviews [22, 87, 88].

Incorporation of Cellular Proteins into HIV-1: Bystanders, Partners, Captives, and Assassins

Viruses by definition heavily depend on host proteins and systems for replication. Unlike DNA viruses which can be quite large and have the genomic capacity to encode a significant amount of synthetic machinery, RNA virus genomes are typically quite small and focused on manipulating the cell with a few proteins to replicate. Examples of this strategy are retroviruses which integrate into the host genome and mimic host genes and picornaviruses which alter the transcriptional machinery to replicate. As discussed above, retroviruses employ many creative expression strategies to maximize viral coding capacity from a small genome. Yet they still rely on host proteins for assistance for all parts of the assembly process.

Cellular proteins are incorporated into and onto HIV-1 virions during assembly. While the functions that cellular proteins provide for the virus inside the cells range from the well studied, e.g., translation, to the less understood, e.g., budding and release, the role, if any for the cellular proteins incorporated into the virion, is less clear. Cellular proteins can associate with the particle in one of four ways: as a consequence of their presence as bystanders present at the site of budding; as partners that assist the viral protein assembly; as captives that are hijacked by specific packaging into HIV to provide a post-assembly function for the virus, e.g., immune evasion or infection; or as assassins, host restriction factors, that have attributes which get them packaged into the virion and inhibit infection. This is a quite large topic and thus is beyond the scope of this review. Interested readers should consult some of the reviews on this subject for details [89, 90]. Each class of incorporated cellular protein provides a helpful clue to the HIV-1 biology. Bystanders, experimentally those proteins that do not require any specific virion protein for incorporation and whose absence does not affect viral replication, mostly implicate the region and nature of the budding site. In contrast, partners, proteins that are specifically incorporated by a virion protein and when absent incorporation the virion fails to assemble, provide mechanistic hints for the cellular side of the assembly process. Captives, experimentally those specifically incorporated proteins in whose absence virions assemble, have reduced infectivity or sensitivity to host antiviral defenses. Unlike the other types, assassins do not play a role in assisting of HIV-1 replication; rather they carryout host suppressive mechanisms; in their absence infectivity and replication are increased, which, in turn, suggest potential approaches for antiviral therapies. Thus, the study of the cellular proteins incorporated into HIV promises to provide clues to the site of budding, assembly, infection and jamming cell defenses, and host restriction factors.

Historically, these proteins were studied using traditional protein chemistry methods. These studies yielded an important but limited set of proteins due to limitations of their sensitivity and ability to examine complex mixtures. Also, some techniques, such as immunoblots, require one to specifically query the sample for a protein of interest rather than identify an unknown protein(s). The evolution of mass spectrometry to provide high sensitivity, high-throughput analysis, and amino acid sequencing of very complex mixtures of proteins has provided for an explosion of proteins associated with HIV-1 virions and provided several important leads for understanding HIV-1 biology [91, 92].

The Purity Problem

In biochemistry, sample purity is paramount to produce unambiguous results. Even though the study of proteins that are incorporated into virions appears to be straightforward, there are several aspects that complicate these studies. When viral proteins are detected in virus particle preparations, it is clear that they are in the virus and not simple contaminants, not so when cellular proteins are found in virion preparations because large amounts of cellular proteins are released into culture media and biological fluids that can contaminate the sample. These proteins are secreted from the cell by normal cellular processes, released by cell death induced by HIV-1, and present in cell culture medium supplements. Therefore, one critical issue is whether a cellular protein detected is truly in/on the particle or simply present as a contaminant in the sample. Retroviral isolation procedures typically use ultracentrifugation to isolate the virions from the extracellular material. Even though it is possible to remove these proteins by biophysical means, by either density or velocity centrifugation, care must be taken to achieve highly effective removal of soluble proteins. In addition to soluble proteins, even uninfected cells release protein-laden vesicles, either microvesicles (particles that bud from the plasma membrane) or exosomes (particles that first bud into late endosomal compartments that, in turn, fuse to the plasma membrane to be released from the cell) [90, 93]. While these vesicles are fairly heterogeneous, a significant subset of these vesicles has the same density and size as HIV-1 particles. Therefore, while biophysical methods can remove a majority of vesicular contamination, they are unable to remove this fraction which contains a large amount of protein. Even more confounding is that the cellular protein makeup of these vesicles roughly mirrors that of virions. This problem is greatest for virus preparations produced from lymphoid cell lines, which produce a large amount of vesicles. But even preparations produced from transfected epithelial cells contain significant quantities of contaminating vesicles. Effective removal can be achieved by supplementing biophysical methods with other approaches.

Two methods currently used to produce highly pure virion preparations exploit differences between the virus and vesicles, protease digestion, and vesicular immunoaffinity depletion. The digestion approach takes advantage of the fact that virion preparations can be digested with high levels of a nonspecific protease, commonly subtilisin or protease K [90, 93]. These proteases remove the proteins on the outside of the virion, but cannot cross the membrane envelope of the virions, leaving the interior proteins intact. In contrast, the protease digests most of the proteins in the contaminating vesicles, making them lighter. Repurifying the digested virions by density allows for effective removal of these particles, leaving highly pure HIV-1 particles with interior proteins intact and ready for study. Another method exploits the incorporation of CD45 into vesicles produced from hematopoietic cells [90, 93]. This highly abundant surface protein is excluded from virions, presumably due to the inability of the large CD45 cytoplasmic tail to fit into the constrained Gag lattice during assembly and budding. Removal of vesicles, which contain CD45, can be accomplished by immunoaffinity depletion with anti-CD45 microbeads. Despite these powerful tools, no method can remove all of the contaminating particles. Therefore, it is important to establish uninfected controls to monitor the efficiency of the removal of potentially contaminating proteins by using parallel-treated mock virus preparations. Even with the most careful purification, it is unreasonable to expect that contaminating proteins in a complex biological sample can be reduced to zero; there will always be a possibility that a protein is not truly on or in the virus. With increased sensitivity, there is increased noise from contaminating particles, potentially leading to false positives. While this is a problem with classical biochemistry methods, it is magnified as greater sensitivity mass spectrometry methods are used. Therefore, conformational assays using other often less sensitive methods and direct methods, e.g., confirming mass spectrometry data with immunoblot or Edman protein sequencing, greatly assist in concluding that a particular protein is inside HIV-1 particles.

One occasionally mentioned effort to support specific incorporation is whether a particular protein is at an "enriched" level in the virus versus that found in the cell. On the surface this seems to be logical; however, the great assumption with this line of reasoning is that proteins are uniformly distributed in a cell, which is clearly not true. In fact, proteins are localized to different regions of the plasma membrane. For instance, one would not expect histones to be present in a virus that buds from the plasma membrane, while finding β-actin would be highly likely. Therefore, enrichment over the total cell or even the plasma membrane does not prove specific incorporation as it is the protein composition at the site of budding that is pertinent. Conversely, the array of bystander proteins incorporated into the virions does provide for sort of fingerprint of where the virus assembled and budded from. Taken one step further, it could be possible to determine the source of a virion by its composition, whether it came from a T cell or macrophage and what type could be inferred by the spectrum of proteins found in it.

HIV-1 Proteome Summary

HIV-1 virions contain a variety of proteins, both viral and cellular. Although there has been much progress made in examining these proteins in the virion, it is important to appreciate that there are many more questions still to be answered: what is the comprehensive picture of the extent and type of HIV-1 protein posttranscriptional modifications? To what extent are virion proteins processed further into minor cleavage sites and do they have any regulatory or replication function for HIV-1? What are the differences in the constellation of cellular proteins in/on HIV-1 particles produced from, e.g., macrophages and subtypes of T cells? What immune regulatory molecules on the surface of HIV-1 might impact in vivo replication and pathogenesis? Finally, to what extent are cellular proteins in the virion contributing to HIV-1 replication and immune evasion? These and many more make for a fruitful playground for the HIV-1 biochemist and mass spectrometrists.

References

1. Rosenberg N, Jolicoeur P. Retroviral pathogenesis. 1997 [cited]. http://www.ncbi.nlm.nih.gov/books/NBK19378/
2. Rosenberg N, Jolicoeur P. Retroviral pathogenesis. In: Coffin J, Hughes S, Varmus H, editors. Retroviruses. Plainview, NY: Cold Spring Harbor Laboratory Press; 1997. p. 475–583.

3. Weiss SH, Goedert JJ, Sarngadharan MG, Bodner AJ, Gallo RC, Blattner WA. Screening test for HTLV-III (AIDS agent) antibodies. Specificity, sensitivity, and applications. JAMA. 1985;253(2):221–5.
4. Barre-Sinoussi F, Chermann JC, Rey F, Nugeyre MT, Chamaret S, Gruest J, et al. Isolation of a T-lymphotropic retrovirus from a patient at risk for acquired immune deficiency syndrome (AIDS). Science. 1983;220(4599):868–71.
5. Yarchoan R, Klecker RW, Weinhold KJ, Markham PD, Lyerly HK, Durack DT, et al. Administration of 3′-azido-3′-deoxythymidine, an inhibitor of HTLV-III/LAV replication, to patients with AIDS or AIDS-related complex. Lancet. 1986;1(8481):575–80.
6. Brook I. Approval of zidovudine (AZT) for acquired immunodeficiency syndrome. A challenge to the medical and pharmaceutical communities. JAMA. 1987;258(11):1517.
7. Wlodawer A, Erickson JW. Structure-based inhibitors of HIV-1 protease. Annu Rev Biochem. 1993;62:543–85.
8. Pear R. Aids blood test to be available in 2 to 6 weeks. The New York Times [cited]. 1985.http://www.nytimes.com/1985/03/03/us/aids-blood-test-to-be-available-in-2-to-6-weeks.html
9. Finzi D, Hermankova M, Pierson T, Carruth LM, Buck C, Chaisson RE, et al. Identification of a reservoir for HIV-1 in patients on highly active antiretroviral therapy. Science. 1997;278(5341):1295–300.
10. Savarino A. A historical sketch of the discovery and development of HIV-1 integrase inhibitors. Expert Opin Investig Drugs. 2006;15(12):1507–22.
11. Garg H, Viard M, Jacobs A, Blumenthal R. Targeting HIV-1 gp41-induced fusion and pathogenesis for anti-viral therapy. Curr Top Med Chem. 2011;11(24):2947–58.
12. Allen DW. Zone electrophoresis of the proteins of avian myeloblastosis virus. Biochim Biophys Acta. 1967;133(1):180–3.
13. Duesberg PH, Robinson HL, Robinson WS, Huebner RJ, Turner HC. Proteins of *Rous sarcoma* virus. Virology. 1968;36(1):73–86.
14. Franker CK, Gruca M. Structural protein of the Friend virion. Virology. 1969;37(3):489–92.
15. Swanstrom R, Wills JW. Synthesis, assembly, and processing of viral proteins. In: Coffin J, Hughes S, Varmus H, editors. Retroviruses. Plainview, NY: Cold Spring Harbor Laboratory Press; 1997. p. 263–334.
16. Swanstrom R, Wills JW. Synthesis, assembly, and processing of viral proteins. [Internet]. 1997 [cited]. http://www.ncbi.nlm.nih.gov/books/NBK19456/
17. Delelis O, Lehmann-Che J, Saib A. Foamy viruses—a world apart. Curr Opin Microbiol. 2004;7(4):400–6.
18. Temin HM, Mizutani S. RNA-dependent DNA polymerase in virions of *Rous sarcoma* virus. Nature. 1970;226(5252):1211–3.
19. Baltimore D. RNA-dependent DNA, polymerase in virions of RNA tumour viruses. Nature. 1970;226(5252):1209–11.
20. Pederson T. Turning a PAGE: the overnight sensation of SDS-polyacrylamide gel electrophoresis. FASEB J. 2008;22(4):949–53.
21. Vogt VM, Simon MN. Mass determination of *Rous sarcoma* virus virions by scanning transmission electron microscopy. J Virol. 1999;73(8):7050–5.
22. Karn J, Stoltzfus CM. Transcriptional and posttranscriptional regulation of HIV-1 gene expression. Cold Spring Harb Perspect Med. 2012;2(2):a006916.
23. Petropoulos C. Retroviral taxonomy, protein structures, sequences, and genetic maps. 1997 [cited]. http://www.ncbi.nlm.nih.gov/books/NBK19417/
24. Petropoulos C. Retroviral taxonomy, protein structures, sequences, and genetic maps. In: Coffin J, Hughes S, Varmus H, editors. Retroviruses. Plainview, NY: Cold Spring Harbor Laboratory Press; 1997. p. 757–805.
25. Malim MH, Emerman M. HIV-1 accessory proteins—ensuring viral survival in a hostile environment. Cell Host Microbe. 2008;3(6):388–98.
26. Trono D. HIV accessory proteins: leading roles for the supporting cast. Cell. 1995;82:189–92.

27. Sundquist WI, Krausslich HG. HIV-1 assembly, budding, and maturation. Cold Spring Harb Perspect Med. 2012;2(7):a006924.
28. Zhu P, Chertova E, Bess Jr J, Lifson JD, Arthur LO, Liu J, et al. Electron tomography analysis of envelope glycoprotein trimers on HIV and simian immunodeficiency virus virions. Proc Natl Acad Sci U S A. 2003;100(26):15812–158127.
29. Ganser-Pornillos BK, Yeager M, Sundquist WI. The structural biology of HIV assembly. Curr Opin Struct Biol. 2008;18(2):203–17.
30. Wang B. Viral factors in non-progression. Front Immunol. 2013;4:355.
31. Muller B, Tessmer U, Schubert U, Krausslich HG. Human immunodeficiency virus type 1 Vpr protein is incorporated into the virion in significantly smaller amounts than gag and is phosphorylated in infected cells. J Virol. 2000;74(20):9727–31.
32. Kondo E, Gottlinger HG. A conserved LXXLF sequence is the major determinant in p6gag required for the incorporation of human immunodeficiency virus type 1 Vpr. J Virol. 1996;70(1):159–64.
33. Sova P, Volsky DJ, Wang L, Chao W. Vif is largely absent from human immunodeficiency virus type 1 mature virions and associates mainly with viral particles containing unprocessed gag. J Virol. 2001;75(12):5504–17.
34. Park J, Morrow CD. Overexpression of the gag-pol precursor from human immunodeficiency virus type 1 proviral genomes results in efficient proteolytic processing in the absence of virion production. J Virol. 1991;65(9):5111–7.
35. Lee SK, Potempa M, Swanstrom R. The choreography of HIV-1 proteolytic processing and virion assembly. J Biol Chem. 2012;287(49):40867–74.
36. Coffin J, Hughes S, Varmus H. Retroviruses. Plainview, NY: Cold Spring Harbor Press; 1997.
37. Carlson LA, Briggs JA, Glass B, Riches JD, Simon MN, Johnson MC, et al. Three-dimensional analysis of budding sites and released virus suggests a revised model for HIV-1 morphogenesis. Cell Host Microbe. 2008;4(6):592–9.
38. Coren LV, Thomas JA, Chertova E, Sowder 2nd RC, Gagliardi TD, Gorelick RJ, et al. Mutational analysis of the C-terminal gag cleavage sites in human immunodeficiency virus type 1. J Virol. 2007;81(18):10047–54.
39. Chukkapalli V, Ono A. Molecular determinants that regulate plasma membrane association of HIV-1 Gag. J Mol Biol. 2011;410(4):512–24.
40. Bell NM, Lever AM. HIV Gag polyprotein: processing and early viral particle assembly. Trends Microbiol. 2013;21(3):136–44.
41. Bukrinsky MI, Haffar OK. HIV-1 nuclear import: in search of a leader. Front Biosci. 1999;4:D772–81.
42. Checkley MA, Luttge BG, Freed EO. HIV-1 envelope glycoprotein biosynthesis, trafficking, and incorporation. J Mol Biol. 2011;410(4):582–608.
43. Tedbury PR, Ablan SD, Freed EO. Global rescue of defects in HIV-1 envelope glycoprotein incorporation: implications for matrix structure. PLoS Pathog. 2013;9(11), e1003739.
44. Ganser-Pornillos BK, Yeager M, Pornillos O. Assembly and architecture of HIV. Adv Exp Med Biol. 2012;726:441–65.
45. Craigie R, Bushman FD. HIV DNA integration. Cold Spring Harb Perspect Med. 2012;2(7):a006890.
46. Hu WS, Hughes SH. HIV-1 reverse transcription. Cold Spring Harb Perspect Med. 2012;2(10):a006882.
47. Sakuragi J. Morphogenesis of the infectious HIV-1 virion. Front Microbiol. 2011;2:242.
48. Adamson CS, Salzwedel K, Freed EO. Virus maturation as a new HIV-1 therapeutic target. Expert Opin Ther Targets. 2009;13(8):895–908.
49. Malim MH, Bieniasz PD. HIV restriction factors and mechanisms of evasion. Cold Spring Harb Perspect Med. 2012;2(5):a006940.
50. Strebel K, Luban J, Jeang KT. Human cellular restriction factors that target HIV-1 replication. BMC Med. 2009;7:48.
51. Towers GJ. The control of viral infection by tripartite motif proteins and cyclophilin A. Retrovirology. 2007;4:40.

52. Sastri J, Campbell EM. Recent insights into the mechanism and consequences of TRIM5alpha retroviral restriction. AIDS Res Hum Retroviruses. 2011;27(3):231–8.
53. Luban J, Cyclophilin A. TRIM5, and resistance to human immunodeficiency virus type 1 infection. J Virol. 2007;81(3):1054–61.
54. Cartier C, Sivard P, Tranchat C, Decimo D, Desgranges C, Boyer V. Identification of three major phosphorylation sites within HIV-1 capsid. Role of phosphorylation during the early steps of infection. J Biol Chem. 1999;274(27):19434–40.
55. Misumi S, Inoue M, Dochi T, Kishimoto N, Hasegawa N, Takamune N, et al. Uncoating of human immunodeficiency virus type 1 requires prolyl isomerase Pin1. J Biol Chem. 2010;285(33):25185–95.
56. Datta SA, Temeselew LG, Crist RM, Soheilian F, Kamata A, Mirro J, et al. On the role of the SP1 domain in HIV-1 particle assembly: a molecular switch? J Virol. 2011;85(9):4111–21.
57. Mirambeau G, Lyonnais S, Gorelick RJ. Features, processing states, and heterologous protein interactions in the modulation of the retroviral nucleocapsid protein function. RNA Biol. 2010;7(6):724–34.
58. Thomas JA, Gorelick RJ. Nucleocapsid protein function in early infection processes. Virus Res. 2008;134(1–2):39–63.
59. Rein A, Datta SA, Jones CP, Musier-Forsyth K. Diverse interactions of retroviral Gag proteins with RNAs. Trends Biochem Sci. 2011;36(7):373–80.
60. Levin JG, Mitra M, Mascarenhas A, Musier-Forsyth K. Role of HIV-1 nucleocapsid protein in HIV-1 reverse transcription. RNA Biol. 2010;7(6):754–74.
61. Meng B, Lever AM. Wrapping up the bad news: HIV assembly and release. Retrovirology. 2013;10:5.
62. Welker R, Hohenberg H, Tessmer U, Huckhagel C, Krausslich HG. Biochemical and structural analysis of isolated mature cores of human immunodeficiency virus type 1. J Virol. 2000;74(3):1168–77.
63. Fujita M, Otsuka M, Nomaguchi M, Adachi A. Multifaceted activity of HIV Vpr/Vpx proteins: the current view of their virological functions. Rev Med Virol. 2010;20(2):68–76.
64. Martin-Serrano J. The role of ubiquitin in retroviral egress. Traffic. 2007;8:1297–303.
65. Ott DE, Coren LV, Sowder II RC, Adams J, Schubert U. Retroviruses have differing requirements for proteasome function in the budding process. J Virol. 2003;77(6):3384–93.
66. Gurer C, Berthoux L, Luban J. Covalent modification of human immunodeficiency virus type 1 p6 by SUMO-1. J Virol. 2005;79(2):910–7.
67. Radestock B, Morales I, Rahman SA, Radau S, Glass B, Zahedi RP, et al. Comprehensive mutational analysis reveals p6Gag phosphorylation to be dispensable for HIV-1 morphogenesis and replication. J Virol. 2013;87(2):724–34.
68. Votteler J, Neumann L, Hahn S, Hahn F, Rauch P, Schmidt K, et al. Highly conserved serine residue 40 in HIV-1 p6 regulates capsid processing and virus core assembly. Retrovirology. 2011;11.
69. Pettit SC, Clemente JC, Jeung JA, Dunn BM, Kaplan AH. Ordered processing of the human immunodeficiency virus type 1 GagPol precursor is influenced by the context of the embedded viral protease. J Virol. 2005;79(16):10601–7.
70. Louis JM, Clore GM, Gronenborn AM. Autoprocessing of HIV-1 protease is tightly coupled to protein folding. Nat Struct Biol. 1999;6(9):868–75.
71. Pettit SC, Simsic J, Loeb DD, Everitt L, Hutchison CAD, Swanstrom R. Analysis of retroviral protease cleavage sites reveals two types of cleavage sites and the structural requirements of the P1 amino acid. J Biol Chem. 1991;266(22):14539–47.
72. Daniels SI, Davis DA, Soule EE, Stahl SJ, Tebbs IR, Wingfield P, et al. The initial step in human immunodeficiency virus type 1 GagProPol processing can be regulated by reversible oxidation. PLoS ONE. 2010;5(10), e13595.
73. Jayappa KD, Ao Z, Yang M, Wang J, Yao X. Identification of critical motifs within HIV-1 integrase required for importin alpha3 interaction and viral cDNA nuclear import. J Mol Biol. 2011;410(5):847–62.
74. Wilen CB, Tilton JC, Doms, RW. HIV: cell binding and entry. Cold Spring Harb Perspect Med. 2012;2(8):pii: a006866.

75. Garg H, Blumenthal R. Role of HIV Gp41 mediated fusion/hemifusion in bystander apoptosis. Cell Mol Life Sci. 2008;65(20):3134–44.
76. Postler TS, Desrosiers RC. The tale of the long tail: the cytoplasmic domain of HIV-1 gp41. J Virol. 2013;87(1):2–15.
77. Chan WE, Lin HH, Chen SS. Wild-type-like viral replication potential of human immunodeficiency virus type 1 envelope mutants lacking palmitoylation signals. J Virol. 2005;79(13): 8374–87.
78. Strebel K. HIV accessory proteins versus host restriction factors. Curr Opin Virol. 2013;3(6): 692–9.
79. Zhao RY, Li G, Bukrinsky MI. Vpr-host interactions during HIV-1 viral life cycle. J Neuroimmune Pharmacol. 2011;6(2):216–29.
80. Fassati A. HIV infection of non-dividing cells: a divisive problem. Retrovirology. 2006;3:74.
81. Koppensteiner H, Brack-Werner R, Schindler M. Macrophages and their relevance in human immunodeficiency virus type I infection. Retrovirology. 2012;9:82.
82. Kogan M, Rappaport J. HIV-1 accessory protein Vpr: relevance in the pathogenesis of HIV and potential for therapeutic intervention. Retrovirology. 2011;8:25.
83. Bukovsky AA, Dorfman T, Weimann A, Gottlinger HG. Nef association with human immunodeficiency virus type 1 virions and cleavage by the viral protease. J Virol. 1997;71(2):1013–8.
84. Foster JL, Denial SJ, Temple BR, Garcia JV. Mechanisms of HIV-1 Nef function and intracellular signaling. J Neuroimmune Pharmacol. 2011;6(2):230–46.
85. Vermeire J, Vanbillemont G, Witkowski W, Verhasselt B. The Nef-infectivity enigma: mechanisms of enhanced lentiviral infection. Curr HIV Res. 2011;9(7):474–89.
86. Seelamgari A, Maddukuri A, Berro R, de la Fuente C, Kehn K, Deng L, et al. Role of viral regulatory and accessory proteins in HIV-1 replication. Front Biosci. 2004;9:2388–413.
87. Rabson AB, Graves BJ. Synthesis and processing of viral RNA. In: Coffin J, Hughes S, Varmus H, editors. Retroviruses. Plainview, NY: Cold Spring Harbor Press; 1997.
88. Rabson AB, Graves BJ. Synthesis and processing of viral RNA. 1997 [cited]. http://www.ncbi.nlm.nih.gov/entrez/query.fcgi?cmd=Retrieve&db=PubMed&dopt=Citation&list_uids=21433339
89. Cantin R, Methot S, Tremblay MJ. Plunder and stowaways: incorporation of cellular proteins by enveloped viruses. J Virol. 2005;79(11):6577–87.
90. Ott DE. Cellular proteins detected in HIV-1. Rev Med Virol. 2008;18:159–75.
91. Linde ME, Colquhoun DR, Mohien CU, Kole T, Aquino V, Cotter R, et al. The conserved set of host proteins incorporated into HIV-1 virions suggests a common egress pathway in multiple cell types. J Proteome Res. 2013;12(5):2045–54.
92. Chertova E, Chertov O, Coren LV, Roser JD, Trubey CM, Bess Jr JW, et al. Proteomic and biochemical analysis of purified human immunodeficiency virus type 1 produced from infected monocyte-derived macrophages. J Virol. 2006;80(18):9039–52.
93. Ott DE. Purification of HIV-1 virions by subtilisin digestion or CD45 immunoaffinity depletion for biochemical studies. Methods Mol Biol. 2009;485:15–25.

Chapter 3
25 Years of HIV-1 Biochemistry

David E. Ott

Retroviral biochemistry came to the forefront in the late 1960s, the co-discovery of reverse transcriptase by Howard Temin's and David Baltimore's laboratories [1, 2]. Retroviruses, originally referred to as RNA tumor viruses, oncoviruses, or oncornaviruses, were being studied for their induction of cancer mostly in avian and murine systems [3, 4]. The term retrovirus first appeared in 1975 [5], reflecting the realization that these RNA viruses go against the "central dogma" of molecular biology, which held that DNA was the principal source of cellular information with genes being expressed by transcription into RNA which is then translated into proteins [6, 7]. RNA tumor viruses were found to reverse this going from RNA to a viral DNA form that is integrated into the host to form a provirus, a mechanism first hypothesized in the 1960s by Howard Temin (see his Nobel Prize lecture [8] for a historical review). The HIV-1 proviral form acts as a stable locus of genes in the cell, producing both viral proteins and the viral RNA genome through a complex series of molecular gymnastics. Hence, the term retrovirus for backward virus was coined. However, it is important to remember that this major advance as well as other in retroviral biochemistry was before the advent of many of the modern biochemical tools, instead relying on rudimentary forms of analysis, protein purification, enzymology, chromatography, protein sequencing [9, 10], spectrometry, velocity sedimentation, density centrifugation, and immunodiffusion. However, progress was slow and the ability to accurately analyze complex mixtures of proteins was nonexistent, unlike today with current mass spectrometry methods.

As discussed in chapter "Introduction: HIV-1 Proteomics, Why Should One Care?", the study of retroviruses has produced an abundant harvest of insights into cancer, immunology, cell biology, antiviral vaccines, biochemistry, and genetics.

D.E. Ott, Ph.D. (✉)
Leidos Biomedical Research, Inc., Frederick National Laboratory
for Cancer Research, 1050 Boyles Street, Frederick, MD 21702, USA
e-mail: ottde@mail.nih.gov

D.R.M. Graham, D.E. Ott (eds.), *HIV-1 Proteomics*,
DOI 10.1007/978-1-4939-6542-7_3

The mechanistic knowledge gleaned from studying the prototypic avian and murine retroviruses laid an essential foundation for the rapid progress in characterizing HIV-1 and AIDS drugs. The early, pre-HIV, study of retroviruses was carried out with fairly basic techniques, basic chromatography, and enzyme assays. Truly these early retrovirologists accomplished amazing things, relying on their brains and careful, well-thought-out experiments more than high-tech techniques. This chapter provides an introduction and discussion of the more modern "classical" techniques which complement the newer "proteomic" mass spectrometry approach and are as essential today as they were decades ago.

Aspects of HIV-1 Particle Biochemistry

Classical biochemistry starts with examining a highly complex mixture of material usually in the form of a tissue from which cells are isolated and lysed, and the components, protein, lipid, or carbohydrate, of interest, are purified and studied. For proteins, this is typically an involved process to isolate the relatively tiny amounts of the protein of interest in the vast sea of cellular proteins. Compared to classical cellular biochemistry, retroviral protein biochemistry is easier since the virus does most of the purification work by releasing particles into the cell culture medium leaving the complexity of the cell behind. By doing so, viruses conveniently purify themselves for the researcher, who can straightforwardly isolate HIV-1 virions using their biophysical properties, either by their density or their size using density or sedimentation centrifugation.

Within the last 15 years, there has been a strong interest in the cellular proteins incorporated into HIV-1 virions [11–13]. Viruses, by definition, require cells to replicate. HIV-1, retroviruses in general, has a relatively small genome when compared to the large DNA viruses which have extensive genomes that code for many proteins which provide independent viral replication functions and modulation cell functions for immune escape and manipulation of the cell for the advantage of the virus. Therefore HIV-1 relies on mostly cellular proteins to replicate. One way to study this is to examine the host proteins found in virions (see chapter "HIV-1 Biology at the Protein Level" for an extended discussion). Host proteins can be incorporated into virions by just being present at the site of HIV-1 budding, being taken up as bystanders due to their presence in the plasma membrane as the particle buds through the membrane. These proteins, while not specifically incorporated, still provide for clues to the site of budding and cell type producing the virus. Proteins could also be incorporated as partners when they interact with viral proteins as well as actively assist in assembly and virion production. Also, HIV-1 could incorporate cellular proteins as captives to assist in post-assembly functions such as immune evasion and promotion of cellular infection.

Thus, while the study of the cellular proteins in virions is important, the potential for contamination is a critical concern because while the origin of the HIV-1 proteins in virions is obvious, great care must be carried out to selectively detect those cellular proteins both in and on the virion versus those that are present as contaminants. The essential question is how to show that a protein is on or in the virion as opposed to being merely being in "purified" virus preparations. For instance, one classical method for surface proteins is to specifically immunoprecipitate the virus and then assay for a viral protein, typically capsid. This is essentially a qualitative method and unable to examine proteins inside the virion. For quantitative studies, the virions must be purified and the particles examined directly. While separation of virions from culture media and most proteins released by cells can be accomplished by using centrifugation, still a significant amount of cellular proteins co-purify with retroviral virions, being present in vesicles (microvesicles or exosomes [12, 14]) that have the same size and density as virions. Thus, studies seeking to identify and characterize cellular proteins in the HIV-1 particles require high levels of purification with strict controls that demonstrate that the protein(s) of interest are removed. Two techniques, protease digestion of virions and CD45 immunoaffinity depletion, are discussed in chapter 2: "HIV-1 Biology at the Protein Level" so they will not be recapitulated here. Nevertheless, in biochemistry, either classical or high-tech mass spectrometry, the purity of the sample is critical to draw accurate conclusions from the results.

HIV-1 Proteomics Before There Was Proteomics

Before the current mass spectrometric sequencing capabilities, analysis of viral and cellular proteins in retroviruses was done with well-developed techniques that relied on classical biochemistry which lack the throughput of proteomics but still identified and characterized the HIV-1 viral proteins and their modifications as wells as many cellular proteins present both in and on HIV-1 particles. In the days before proteomics, analysis of proteins in the virion took a brute-force approach (techniques are discussed below): isolate the various proteins in the virion by chromatography, run the fractions on a protein gel, blot the proteins on a gel, and cut out bands for automated microsequencing. This approach was labor intensive and took months/years to complete. In contrast, this can be done in a couple of days with much higher sensitivity with current MS/MS spectrometry. Yet the classical approach and its associated techniques still have some advantages. The following is a review of these classical biochemistry techniques, their strengths and limitations.

Sodium Dodecyl Sulfate Polyacrylamide Gel Electrophoresis

An essential advance in biochemistry arrived with the advent of sodium dodecyl sulfate polyacrylamide gel electrophoresis (SDS-PAGE), a critical technique that is widely taken for granted today. Before SDS-PAGE proteins were electrophoresed on a variety of gel substrates and buffer systems. Because proteins can have neutral, positive, or negative charges, the samples were loaded in the center of the gel and electrophoresed with proteins migrating according to their charge, positive to the cathode side and negative to the anode. Thus, while proteins could be separated using this system, relative size was not determined. Note that this was not isoelectric focusing which yields the protein's isoelectric point, a measure of the protein's pH properties [15]. This changed in 1966 with the first use of the negatively charged detergent sodium dodecyl sulfate (SDS) in the sample which was electrophoresed in thin polyacrylamide tube gels (see references in [16]). Since the SDS not only solubilizes most proteins, but imparts a net negative charge to the protein, the proteins migrate in only one direction, toward the anode. Thus, proteins could be loaded on the top of the gel and unidirectionally electrophoresed down the gel. It was soon after the introduction of SDS-PAGE (1967) that it was recognized that the SDS bound most proteins according to their mass. Thus, amount of negative charge imparted to the proteins by SDS is, for the most part, proportional to the molecular mass of the protein [16], a major breakthrough. However, at the time, polyacrylamide was cast in tube gels which meant every sample had to be run in an independent tube, making comparison of samples difficult. Now samples with different proteins could be meaningfully compared based on a consistent measurable property, their mass. The final major step was the development of the slab-gel SDS-PAGE system by Laemmli in 1970 [17] which involved casting a polyacrylamide gel as a rectangular slab using a Trizma HCl (Tris)-glycine-SDS buffer system and a discontinuous gel system that used a small "stacking" gel on top the analytical gel to focus the protein sample before it entered the main gel. This system now allowed a wide variety of protein samples to be resolved side by side with a molecular mass marker for accurate comparison and determination of their apparent molecular weight. While later innovations such as gels with increasing gel concentrations, gradient gels, and alternative buffer systems have appeared, the Laemmli system remains the basis for analyzing proteins, and the Laemmli paper is one of the most cited in biological research [17].

Immunoblotting/Western Blotting

The discovery of the DNA code opened a host of new ways to analyze DNA, a molecule once thought of as a gooey, stringy nuisance contaminant by protein biochemists. One of these was the Southern blot, eponymously named for its inventor presented in another seminal paper [18, 19]. For this technique, DNA was separated on an agarose gel, blotted onto a sheet on nitrocellulose, and then hybridized with specific radiolabeled DNA probes to reveal the presence of the specific DNA in the mixture. This watershed advance in nucleic acid research was soon followed by applying this technique to the analysis of RNA (Northern) and protein (Western blot or immunoblot). For the Western blot, SDS-PAGE gels are electroblotted onto polyvinylidene fluoride (PVDF) membranes and reacted with serum or antibodies specific for a protein or proteins. The bound antibodies are visualized by various means, having been labeled directly with a fluorophore, enzyme, or radioisotope before the analysis or detected secondarily using a labeled anti-immunoglobulin antibody which binds to the primary antibody. This technique brought even more analytical power to SDS-PAGE techniques by being able to identify specific proteins as immunoreactive bands in the sample lane, no matter how complex the mixture.

Western blots are an essential tool for the analysis of proteins and are widely used by HIV researchers. There are primarily two types of immune reagents used to visualize proteins of interest, antiserum or monoclonal antibody. The key to the success of this technique is the strength and specificity of these reagents, each of which has its distinct advantages. Antiserum (polyclonal sera) made to the whole or just a portion of the protein of interest commonly recognizes more than one epitope and thus can detect more than one portion of the protein. Also, these immune reagents can be stronger due to dominant epitope specificities in serum. For instance, we have some antisera that readily detect viral proteins at a 1:500,000 dilution for analysis compared to a typical dilution of 1:4000 for antisera and 1:1000 for monoclonal antibodies. However, this is not always the case as there are plenty of weak antisera and strong monoclonal antibody titers. Additionally, because there can be an array of antigen-specific antibodies present, antisera can recognize several epitopes within a protein; Western blots using antisera are much more plastic in detecting proteins. Therefore, variant proteins such as mutants, isotypes, splice variants, related proteins, and multiple fragments of proteolytically processed proteins may be detected. Antiserum made to short defined peptides do not have these advantages since they typically elicit only one epitope specificity. A disadvantage to antisera is that it contains not only the desired antibodies, but the other antibodies in present in the serum, so it is possible that these reagents could produce higher background noise and nonspecific bands. For instance, many antisera are produced by linking a peptide to bovine serum albumin which is also a component of cell culture media and commonly used in molecular mass standards. Thus, these types of peptide antisera can lead to strong unwanted bands in blots that can overwhelm and obscure bands from the desired proteins. Finally, the supply of any given antiserum is finite and cannot be precisely reproduced.

Monoclonal antibodies recognize only one epitope, thus only one part of the protein. Because the epitope is usually defined, being made to a peptide, the interpretation of the results is more straightforward. This can be useful for mapping proteolytic fragments or splice variants with two or more antibodies. Also, a monoclonal antibody can have less background, having only one antibody present instead of the many specific and much more nonspecific antibodies present in the serum. Despite having only one specificity, there are many monoclonal antibodies that detect one or more irrelevant bands, due to the antibody cross-reacting with similar epitope on another protein. Finally, monoclonal antibodies are produced from hybridoma cell lines either in vitro or in vivo, so a good antibody can be reproduced. Also, due to their purity, monoclonal antibodies are commonly directly labeled for detection of proteins and as secondary detection reagents in Western blots.

Despite the analytical power and sensitivity of current mass spectrometry for analyzing HIV-1, both SDS-PAGE and its logical extension, the Western blot, still complement modern "proteomic" techniques. First, they are rapid and easily interpretable, with some procedures giving results within a couple of hours with low cost. The instrumentation is comparatively simple so each laboratory member can have their own equipment and carry out analyses independently. Most importantly, these techniques examine the whole protein, not just the sequence of peptide fragments by MS/MS spectrometry in complex mixtures of proteins which leads to ambiguity: was this sequence from a pro-form of a protein, a processed active form, or a variant, or, in the case of the Gag polyprotein, is the peptide from Gag or just one of its mature proteins? Thus, many researchers confirm proteomic protein identifications with a Western blot. Alternatively, the complexity of a mass spectrometry sequencing sample can be reduced by fractionating by size. The sample can be separated on an SDS-PAGE gel which is the cut into size fractions and the proteins in each slice digested and analyzed to provide the sequences present in a specific molecular mass range. An Achilles heel of mass spectrometry is the bioinformatics side where the peptide sequences are matched to proteins. Because peptide sequences are compared to expected values in databases, genetic polymorphisms, unexpected protein modifications, and other factors can cause vital peptide to be missed obscuring the identification of proteins. Finally, these classic biochemistry and proteomic techniques have opposing, thus complementary natures. Western blots look inward: one needs to decide what protein to interrogate the blot for a principally yes or no answer. Since these antibodies need to be chosen, one needs to test from a hypothesis of what protein is of interest as it is obviously not profitable to guess with tens of thousands of proteins in a cell. In contrast, MS/MS spectrometry sequencing is outward looking, one harvests information on nearly all of the proteins in a complex mixture simultaneously, yet one does not study the intact protein. Mass spectrometry and Western blots work hand in hand, proteomics indentifying candidates and blots providing valuable information about the form of the protein and its posttranslational modifications.

Reversed-Phase High-Pressure Chromatography

Chromatography in many different forms has been used for over a century to separate materials. Biochemically, paper, silica plates, and columns (mostly liquid in biology) with various packing materials have been used for years to separate tissue and cell extracts into pure proteins. Four principle types of liquid chromatography column are molecular sieve/gel exclusion which separates on the basis of size; ion exchange which exploits differences in charge at a pH, i.e., their isoelectric point (pI); affinity which relies on binding of a protein to a material, DNA, cofactor, protein, or specific antibody; and reversed phase which separates proteins based on the hydrophobic character. This last method employs a hydrophobic column material through which sample in a hydrophilic solvent is pumped. Since all proteins have some hydrophobic character, most bind to the column packing matrix. The proteins are then separated by increasing the hydrophobicity of the solvent flowing over the column in a gradient. Proteins are eluted from the nonpolar matrix when they partition into the solvent according to their hydrophobicity, i.e., they are more soluble in the solvent than on the matrix. This process is carried out at high pressure to produce a rapid flow rate that is precisely controlled so that the proteins eluted off the column are resolved in narrow peaks at distinct and reproducible times. The primary benefits of reversed-phase high-pressure liquid chromatography (r-pHPLC) are high resolution, speed, and reproducibility; many virion proteins can be purified in one column run. The r-pHPLC technique is better than ionic exchange because the hydrophobic nature of a protein is typically little changed, whereas a protein charge can vary widely based on oxidation, reduction, and differential posttranslational modification of the proteins. Molecular exclusion columns have relatively low-resolution power compared to r-HPLC so they can rarely resolve proteins out of complex mixtures. One of the early downsides to using r-HPLC is that the column sizes were relatively large, therefore requiring large amounts of sample. This is good for preparative work, for separating proteins from milligrams of purified virus, concentrated from liters of virus production, but not for the analytical microgram to tens of nanogram amounts produced from cell transfections or material isolated directly from primary sources. The advent of microbore HPLC techniques, using ~2 mm columns with single digit micron-sized column material, now allows for small amounts of virus (from >1 to 0.1 μg) to be rapidly analyzed [20]. With high precision pumps and programmable gradient controllers, r-HPLC can resolve both the viral and host proteins in the virion at reproducible times allowing for matching of chromatograms by overlaying the profiles to compare several samples. Also, when using a UV detector in the near UV range (206 nm), where the absorbance of amino acids is the same, the peak area is directly proportional to mass, allowing for comparative measurements between protein peaks in the chromatograph.

The downsides to r-HPLC are that this technique requires that the proteins be denatured. Thus, the isolation of active enzymes or intact protein complexes such as multiprotein complexes is not possible, unless they can be refolded/reassociated post-purification. Another drawback is that r-HPLC samples cannot contain detergent, a common technique for producing cell extracts, as this binds to the column and interferes with the partitioning of the solute proteins between the column and the solvent. While both of these limit r-HPLC utility for analysis of cells, the goal of analyzing retrovirus particles is to examine their protein composition rather than enzymology. Also since the complexity of proteins in virions is relatively simple compared to cells, they can be lysed directly in the r-HPLC loading buffer without extraction and isolation procedures involving detergent. Despite these limitations, r-pHPLC remains a powerful technique for the examination of both viral and cellular proteins in HIV-1, simultaneously analyzing and isolating the components of virions.

Automated N-Terminal (Edman Degradation) Protein Sequencing

Long before DNA could be sequenced [21], proteins were being routinely sequenced by using chemicals or enzymes. The original Sanger reagent sequencing, which chemically labeled the amino-terminal residue from a polypeptide which was then partially hydrolyzed and the fragments separated and completely hydrolyzed

(1951), was soon supplanted by the Edman degradation technique which immobilized the carboxy-terminus of the protein and then chemically cleaved the amino-terminal residue so that each step revealed a successive amino acid [9, 10, 17]. These techniques required large amounts of protein and were carried out painstakingly by hand. Automating this process, the protein sequenator developed by Edman in 1967 [22] made protein sequencing more rapid but still required large amounts of protein and the length of sequence that could be read was rather limited. Therefore, to sequence proteins, one had to break them up into small polypeptides by cleaving the protein with proteases and then sequencing the peptides. By manually comparing and knitting together the short sequences produced from several protease digests, a sequence of the whole protein could be obtained. Modern protein sequencing machines use an approach pioneered by Hunkapiller and Hood, who developed the protein microsequencer in 1978 [23]. With this new tool, it was possible to sequence 77 amino acids from using 5 mg of antibody light chain. Much longer sequence reads from much less material in an automated format revolutionized protein sequencing, vastly increasing the speed and ease of sequencing.

Today, the chemically based microsequencer is still used. Its disadvantages are that it requires much more protein than ms/ms spectrometry sequencing. Also, it works best on a purified protein; the presence of other proteins causes other amino acids to be present in the reads. While low levels of contaminants can be tolerated, being less abundant than the proteins of interest, not all amino acids yield the same signal; thus, it is possible to have a weakly yielding residue in the protein at the same position as a strong amino acid signal in the contaminant which could lead to a miscall. Thus, protein sequencers cannot analyze complex mixtures of proteins. Also, the chemistry of Edman degradation sequencing requires a free amino group on the end. Therefore, proteins with modified amino-termini, e.g., those with acyl modifications such as HIV-1 Gag, cannot be directly sequenced [23]. To sequence the remainder of the protein, this modified amino acid must be removed, typically by a downstream protease cleavage to produce a "free" amino-terminal end. This is a real shortcoming as roughly half of the proteins in the cell have blocked amino-termini. Furthermore, "reading" the sequence output requires some skill that must be learned from experience: how to adjust the results for differing amino acid yields and looking for potentially modified amino acids, e.g., ubiquitinated proteins or cyclic proteins, which make interpretation difficult. On the positive side in comparison to MS/MS spectrometry sequencing, the length of the sequencing reads is much longer and is direct. The sequence is obtained without any bioinformatical statistical fitting to a protein sequence database that could miss proteins due to allelic polymorphism, mutations, or posttranslational modifications not noted in the database. Also, some proteins are cleaved into too small a polypeptide to be statistically significant enough to be reported. This is especially true with basic proteins and trypsin digestion, a commonly used protease for fragmentation. Finally, the coverage of an MS/MS spectrometry sequence is only as good as its database and its digestion procedure. Despite its low throughput and other shortcomings, the protein microsequencer still remains an important protein analysis tool.

Amino Acid Analysis

Amino acid analysis analyzes the amino acids present in a protein by first acid hydrolyzing the protein and then separating and detecting the individual amino acids. This technique, once done manually with thin layer chromatography, is now carried out in an automated format which greatly increases speed and accuracy. This method provides the most precise way to analyze the amino acid makeup of a protein. Amino acid analysis also is the most accurate method to measure the absolute amount of protein in a sample. However, a downside to this technique is that it requires relative larger amounts of proteins than some other methods and samples that are very pure. In HIV-1 biochemistry, this was used to identify the amino acid composition of the HIV virion proteins before sequencing. With the roster of amino acids found in a protein, the sequencing is more easy to read and assemble since one only needs to arrange the amino acids like a puzzle rather than discover and then order them. Sequence ambiguities such as missing amino acids and miscalls are more easily resolved. Also, once the sequence is known, then modified residues can be identified by their differing elution profile on the HPLC separation. This technique is also useful for quality control of recombinant and purified proteins.

These classical techniques have served us well over the last few decades and are still very relevant to the study of proteins in HIV-1 particles. Looking to the future, there will be a long partnership between the old-school "proteomics before there was proteomics" and proteomics.

Acknowledgments This project has been funded in whole or in part with Federal funds from the National Cancer Institute, National Institutes of Health, under Contract No. HHSN261200800001E. The content of this publication does not necessarily reflect the views or policies of the Department of Health and Human Services; nor does the mention of trade names, commercial products, or organizations imply endorsement by the US government.

References

1. Baltimore D. RNA-dependent DNA, polymerase in virions of RNA tumour viruses. Nature. 1970;226(5252):1209–11.
2. Temin HM, Mizutani S. RNA-dependent DNA polymerase in virions of *Rous sarcoma* virus. Nature. 1970;226(5252):1211–3.
3. Vogt P. Historical introduction to the general properties of retroviruses. In: Coffin J, Hughes S, Varmus H, editors. Retroviruses. Plainview, NY: Cold Spring Harbor Laboratory Press; 1997. p. 1–26.
4. Vogt P. Historical introduction to the general properties of retroviruses. Retroviruses 1997. [cited]. http://www.ncbi.nlm.nih.gov/books/NBK19388/
5. Online M-W. 2014. [cited]. http://www.merriam-webster.com/dictionary/retrovirus
6. Crick FH. On protein synthesis. Symp Soc Exp Biol. 1958;12:138–63.
7. Crick F. Central dogma of molecular biology. Nature. 1970;227(5258):561–3.
8. Temin HM. The DNA, provirus hypothesis. Science. 1976;192(4244):1075–80.

9. Sanger F. Chemistry of insulin; determination of the structure of insulin opens the way to greater understanding of life processes. Science. 1959;129(3359):1340–4.
10. Edman P. On the mechanism of the phenyl isothiocyanate degradation of peptides. Acta Chem Scand. 1956;10:761–8.
11. Ott DE. Potential roles of cellular proteins in HIV-1. Rev Med Virol. 2002;12(6):359–74.
12. Ott DE. Cellular proteins detected in HIV-1. Rev Med Virol. 2008;18:159–75.
13. Chertova E, Chertov O, Coren LV, Roser JD, Trubey CM, Bess Jr JW, et al. Proteomic and biochemical analysis of purified human immunodeficiency virus type 1 produced from infected monocyte-derived macrophages. J Virol. 2006;80(18):9039–52.
14. Coren LV, Shatzer T, Ott DE. CD45 immunoaffinity depletion of vesicles from Jurkat T cells demonstrates that "exosomes" contain CD45: no evidence for a distinct exosome/HIV-1 budding pathway. Retrovirology. 2008;5(1):64.
15. Wikipedia. Isoelectic point. [cited]. http://en.wikipedia.org/wiki/Isoelectric_point
16. Shapiro AL, Vinuela E, Maizel Jr JV. Molecular weight estimation of polypeptide chains by electrophoresis in SDS-polyacrylamide gels. Biochem Biophys Res Commun. 1967;28(5):815–20.
17. Laemmli UK. Cleavage of structural proteins during the assembly of the head of bacteriophage T4. Nature. 1970;227(5259):680–5.
18. Southern EM. Detection of specific sequences among DNA fragments separated by gel electrophoresis, 1975. Biotechnology. 1992;24:122–39.
19. Soutnern E. Detection of specific sequences among DNA fragments separated by gel electrophoresis. J Mol Biol. 1975;98:503–71.
20. Scott RP, Kucera P. Use of microbore columns for the separation of substances of biological origin. J Chromatogr. 1979;185:27–41.
21. Donelson JE, Wu R. Nucleotide sequence analysis of deoxyribonucleic acid. VI. Determination of 3′-terminal dinucleotide sequences of several species of duplex deoxyribonucleic acid using *Escherichia coli* deoxyribonucleic acid polymerase I. J Biol Chem. 1972;247(14):4654–60.
22. Edman P, Begg G. A protein sequenator. Eur J Biochem. 1967;1(1):80–91.
23. Hunkapiller MW, Hood LE. Direct microsequence analysis of polypeptides using an improved sequenator, a nonprotein carrier (polybrene), and high pressure liquid chromatography. Biochemistry. 1978;17(11):2124–33.

Part 2
Modern HIV-1 Proteomics

Chapter 4
Proteomic Studies of HIV-1

David R.M. Graham

Introduction

Up to this point, the reader should have a good understanding of HIV-1 at the protein level. In Chaps. 1 and 2, we have explored in depth the role of host proteins and the techniques that are routinely used in traditional research approaches. In this chapter we will introduce some of the fundamentals of proteomics and how they can best be applied to the study of HIV-1 and other viruses.

When our group started working on HIV-1 proteomics over a decade ago, we had to overcome many challenges. Given the limited amount of sample available for HIV-1, we didn't have enough material to follow stringent process testing like we normally would have done for our classical biochemical experiments. Eventually, with the help of our colleagues at the AIDS and Cancer Vaccine Program, we began using large quantities of 1000-fold concentrated virus isolated from >500 mL of cell culture supernatants. Armed with sufficient virus to systematically address where we were making mistakes, we were able to identify and adapt our methods to work with HIV-1 and began to disarm many experimental and methodological landmines. In this chapter, we share the lessons and experiences that we had to overcome to finally be able to move forward to obtaining meaningful results and publications. It is my hope that the reader will be able to benefit from these experiences as a starting point for the study of HIV-1 or other projects involving limited numbers of samples and specialized approaches.

D.R.M. Graham, M.Sc., Ph.D. (✉)
Department of Molecular and Comparative Pathobiology, Johns Hopkins School of Medicine, 733 N. Broadway, MRB 835, Baltimore, MD 21205, USA
e-mail: dgraham@jhmi.edu

D.R.M. Graham, D.E. Ott (eds.), *HIV-1 Proteomics*,
DOI 10.1007/978-1-4939-6542-7_4

Proteomic Studies of HIV-1

Proteomics: "A Three-Legged Stool"

Proteomics has been described by many as a "three-legged stool"—if any one leg fails, there are catastrophic results for the user. In terms of HIV-1 proteomics, the legs of the stool consist of (1) sample preparation, (2) mass spectrometry, and (3) bioinformatics. In this chapter, we will principally address the first two legs, and we dedicate Chap. 6 to addressing the third. I have attempted to provide practical advice and the necessary minimal amount of background information for investigators wishing to attempt studies in this area or for physicians to better understand the type of information that one might consume as this field matures.

Sample Preparation

The purity problem of HIV-1 was explored in Chap. 2; however, we will expand upon this issue as it applies to proteomics. Ott and colleagues have nicely described the use of both subtilisin and CD45 depletion for the generation of pure virus preparations for host and viral proteins within the lipid bilayer of the virion and whole virions, respectively [1]. Our group has also contributed to this area publishing methods of virus isolation including density manipulation and refined protocols for the use of OptiPrep reagents [2]. As we get closer to completing the initial "cataloging" of host proteins that are incorporated into virions, we are very likely to shift toward examining posttranslational modifications (PTMs) of viral proteins (described in Chap. 5) and the analysis of HIV-1 virions isolated directly from patients. Also, in the near future, we are likely to begin to take more quantitative approaches to the study of HIV-1 virions in patients. This shift in applications also necessitates a shift in our sample preparation procedures—away from the traditional large-scale biochemical approaches typically used in proteomics to those that allow for the more rapid isolation of virions such as affinity purification approaches. For affinity purification approaches, despite being marketed as an HIV-1 purification kit, we only just recently confirmed that CD44 was incorporated into HIV-1 virions from both macrophages and T-lymphocytes [2]. Thus, consideration of a CD44-positive enrichment kit (Miltenyi), as a first step alongside a CD45 depletion kit, may accelerate sample processing. The caveats with these approaches, though, are that CD44 enrichment might preferentially enrich for R5-tropic virions [3] or may miss subsets of viruses that may not incorporate CD44. Although affinity enrichment and depletion methods are very straightforward, ultracentrifugation remains one of the best approaches to isolate HIV-1 [3] and is widely used for the enrichment of virus for ultralow limits of detection by quantitative polymerase chain reaction (PCR) methods. Adaptation of these protocols to the latest generation of benchtop ultracentrifuges that can rapidly obtain extremely high relative centrifugation forces in a short period of time may be an alternative path moving forward.

For the near future, it is likely that primary viruses must be expanded from cell cultures ex vivo unless working with patient samples isolated from patients in the acute phase HIV-1 infection, where HIV-1 titers are very high. If ex vivo expansion of virus is the case, then careful consideration must be given to the expansion strategy, as from a host-protein standpoint, HIV-1 will reflect the cell type that it last replicated in, potentially altering any host-protein phenotypes of PTMs that would have existed in the patient [2]. As tissue culture approaches improve for ex vivo expansion of cells [4] and our knowledge grows about what the differences are between primary virions and those expanded ex vivo, this may become less of an issue in the future. From a sequence perspective, if one is using proteomic approaches to confirm virus sequence changes ex vivo, expansion is less of a concern.

Regardless of the technique for the purification or generation of HIV-1 virions for study, for the purpose of cataloging viral proteins or the study of viral protein PTMs, virus must be significantly concentrated so that total protein concentration is in the range of 1 mg/mL to facilitate both the digestion of virus by proteases (most commonly trypsin) and the cleanup and recovery of the digested material. To this end, ultracentrifugation offers a very rapid way of "pelleting" virus to concentrate it, and the supernatant can be discarded. Even if the resulting pellet is invisible to the naked eye, the virus can then easily be resuspended in a digestion buffer containing trypsin to improve recovery. In our group we use an acid-cleavable detergent like RapiGest (Waters) to accelerate the digestion of the virus and also inactivate the virion at the same time. Acid-cleavable or similar detergents are a must as common detergents are contaminants for mass spectrometry analysis in following steps (reviewed in [5]).

Protein and Peptide Quantitation

While there are a plethora of different protein quantitation kits on the market based upon different methodologies (reviewed in [6]), users are often unaware of the effect of interfering substances, and protein quantitation can be vastly impacted. In our experience, protein and peptide quantitation is one of the most important steps in proteomics. If sufficient experimental material exists (>50 μg), our first recommendation is a precipitation-based approach, like the 2D Quant kit from GE healthcare. Specifically made to quantitate proteins from samples prepared with complex lysis buffers that include detergents, this kit simply precipitates out the protein and then follows a colorimetric approach. The downsides of this approach are the relatively high amount of protein required for this purpose, only really allowing for studies involving infected tissues or cells in the context of HIV-1. Therefore, as an alternative we have used a custom protein binding assay described in 6 with a fluorescent post-dye like Sypro Ruby (Invitrogen) or Lava Purple (epicocconone, the Gel Company/GE Healthcare). This method uses a dot-blot apparatus, and interfering substances are washed through a filter, whereas the protein itself adsorbs to a nitrocellulose membrane. The sample is run along with a standard curve and read on a fluorescent scanner. Protocols for each of these methods are available on the vendor sites. Recently, Feist and Hummon reviewed similar approaches for microgram amounts of materials for proteomic studies [5, 7].

Prior to sample concentration, for most individuals working with HIV-1 ELISAs or quantitative PCR generally is used to determine the quantity of virus present in the sample. This information can be used to help estimate the total amount of protein in a sample. Based upon published studies, there are roughly 1,500 copies of capsid (p24) per copy of viral RNA [8], which works out to be approximately 16,000 virions per picogram of p24. Using results obtained by Marozsan and colleagues, these numbers appear to overestimate the number of virions by a factor ranging from 33 to over 100, when calculating the ratios of p24 to viral RNA [9]. Given the higher reproducibility of viral RNA measures versus p24 assay results, we prefer to normalize by viral copy number. By ELISA, in our experience, [9] the protein concentration of a virus preparation is generally 10–20-fold higher than the p24 level. Like any quantitative assay, one must be very careful to run the sample in dilution to make sure that the reported value is in range of the standard curve used for the assay. In practical terms, we generally use 25 mL of infected culture supernatant to obtain sufficient virus in the supernatant to perform a discovery experiment. Our general assumptions are that HIV-1 grows at approximately 10 ng/mL of p24 from permissive cell lines. Thus, we have 250 ng of p24, or using our rule of thumb, 2.5 μg of total protein. This is on our low end for a thorough cataloging of virus by mass spectrometry but sufficient for us to obtain significant coverage of the virus for most applications.

For discovery experiments, the limited abundance of HIV-1, especially from small cell culture experiments or from primary isolates, is the biggest risk factor for the success in HIV-1 proteomics. A good rule of thumb is to work backward from your optimal analysis to inform yourself of how much material is required for your mass spectrometry experiment. For instance, if one is looking to perform an experiment to find as many proteins as possible in their HIV-1 isolate, then the ideal amount of protein to have for nano-high-performance liquid chromatography (nano-HPLC) ranges between 1 and 10 μg of digested and cleaned peptide on column. Working backward, using the optimal workflow recommended below, always underestimates protein amount after digestion by a factor of 2 prior to digestion of protein and peptide cleanup and quantitation by HPLC. Thus, for a discovery experiment, the risks expand greatly for starting amounts of material under 20 μg. Since purified virus is roughly 1 μg/mL of total protein in supernatant, this means that 20–25 mL of supernatant is the minimum required. For plasma from a patient, this would depend on the titer of virus in the patient, but 20 mL of plasma would be a good starting point for a discovery experiment, so long as the patient is viremic.

Sample Cleanup: Beware of Traditional Detergents

Most classical methods working with HIV-1 involve inactivation of virus with 1 % triton X-100. Since detergents are a profound interfering substance for both nano-HPLC and mass spectrometry, different strategies must be thought of for inactivation of virus. While new approaches and products allow for the removal of detergents [10], many of which are now commercially available (Pierce or Thermo Scientific) in our laboratory, we have substituted the use of an acid-cleavable detergent (reviewed in [11]) such as 0.5 % RapiGest (Waters) or alternatively a buffer containing urea to facilitate digestion of proteins [2] obviating the need to use detergents at all. The use of a high-quality trypsin is also essential following the manufacturer's instructions to avoid contamination with trypsin autolysis products.

The next step of good sample preparation is removing salts (another interfering substance) from the sample and actually quantifying the peptide that will be injected onto the mass spectrometer. In our group we have moved directly to off-line HPLC cleanup of the peptides and use a standard curve by HPLC. By far this is the gold standard for peptide quantitation and desalting, but from a practical standpoint, this is somewhat like driving around the block in a Ferrari, as using HPLC instrumentation for this purpose can be costly and using a neighbor's HPLC system is not practical since often they are set up with assays specific to their laboratory needs. A much less expensive approach is the use of peptide spin cartridges, which allow for washing and desalting of peptides in a typical laboratory setting. The subsequent use of the LavaPep [6] (The Gel Company) or a similar peptide quantitation method, like nanodrop (Thermo Scientific), provides the investigator with precise information on the abundance of peptides prior to performing a mass spectrometry experiment as the recovery can sometimes be variable between spin cartridges. The use of a peptide standard mix to make a standard curve is essential to ensure accurate quantitation (available from ProteoChem, Life Technologies and Sigma-Aldrich). Regardless of the approach, purified quantitated peptides maximize the probability of success of a mass spectrometry experiment.

Mass Spectrometry: The Basics

Prior to discussing the acquisition of mass spectrometry data, a lengthy aside is necessary to introduce mass spectrometry (MS), as the success of the HIV-1 proteomic experiment is now dependent on a well-executed MS experiment. While the end point of a mass spectrometry experiment is obtaining information on both protein composition and abundance, understanding the principles of mass spectrometry (MS) helps inform the investigator as to what types of mass spectrometers are better suited to what purpose. While any experienced proteomic core director should be able to guide the newly initiated, it never hurts to come in somewhat informed to the conversation. Unfortunately, there is a paucity of reviews in mass spectrometry for the lay audience and for the most part a lack of fundamental education for most scientists and clinicians in this area. At the time of writing, the Agard lab at University of California San Francisco (UCSF) has a fantastic primer available online (http://www.msg.ucsf.edu/agard/protocols.html—MS101; Google Keyword search: Agard lab mass spectrometry 101) that I have drawn heavily from my own teaching of the fundamentals of mass spectrometry at Johns Hopkins School of Medicine (https://jh.box.com/ms-basics-graham). The next section presents a lay view of mass spectrometry meant as a general guide for the reader and is in no way meant to be a comprehensive review of the subject but is intended to allow the reader to have an informed conversation with a mass spectrometrist.

Vacuum System and Source

Often the first thing that somebody notices when a mass spectrometer is installed in the laboratory is that they are loud. The noise associated with a MS instrument is due to the vacuum systems needed to keep the instrument operating under very low vacuum ranging from 10^{-3} to 10^{-5} Torr or lower depending on the section of the instrument or what operations are being performed (for reference, outer space pressure ranges from 10^{-6} to $<10^{-17}$ Torr in interstellar space, the moon surface atmosphere being 10^{-11} Torr). Think of the classic experiment that is performed in high school labs across the nation—a falling ball or a falling feather. In normal atmospheric pressure air (760 Torr), a feather falls much more slowly than the ball due to air resistance. In a vacuum they fall at the same rate (see the Human Universe: Episode 4 on Youtube by Brian Cox). Ions, despite their incredibly small sizes, bang into other molecules and are slowed down just like any other matter. Therefore, it's the job of the source region to allow samples to enter into the MS system from an area of high pressure to inside of the instrument—where a very high vacuum exists. The trick with mass spectrometry has always been trying to get the sample from the solid or liquid phase into the gas phase so that it can be moved around inside a vacuum. The other problem is how we can move it around once we have converted our analyte into the gas. The job of the source of a mass spectrometer is to convert our analyte into the gas phase and, at the same time, impart a charge on analyte. The charge, either positive or negative, is the only way we can move something around in the gas phase using principles of magnetism.

The source most widely used today is what is called electrospray ionization mass spectrometry. In this method peptides are resuspended into an aqueous solution, ran through a capillary column (from an HPLC system) through an emitter needle that is electrified at high voltage, while gas, typically nitrogen, is blown into the source region to evaporate the solvent. The result is that the solution, which contains charged particles, rapidly evaporates and the droplets begin to reach a point where the like charges repel each other and the force is stronger than the surface tension of the droplet resulting (Rayleigh limit) in an explosion of the particles out of solution into the gas phase (into the air). Since the most commonly used protease, trypsin, cleaves after a lysine or arginine, which are basic amino acids, peptide typically becomes protonated (H+), resulting in peptides having at least one positive charge. Protonation is facilitated by acidic pH conditions. After being charged into ions, the peptides are focused through a series of ion optics toward the next component of the system, the mass analyzer. While air and other non-charged particles are also entering into the instrument, the ion optics create fields that are stronger than the airflow created by the vacuum system. Therefore, the charged particles, or ions, are continually concentrated relative to their environment as they enter into regions of the instrument held at lower pressure.

The Mass Analyzer

As mentioned above, vacuum is going to allow charged peptides to move in the instrument. Highly efficient (roughing) and high velocity (turbo pumps) vacuum pumps are used to remove as much air as is feasible to create a vacuum. Once this is achieved, peptides can be accelerated, decelerated, and steered using magnetic fields. While beyond the scope of this book, in most Electrospray Ionization (ESI) instruments, a series of ion optics are used to steer the beam of ions and focus them to the mass analyzer where masses are separated.

The job of the mass analyzer is to separate different masses entering through the source regions. The same principles that are involved in redirecting ions (varying voltages and radio frequencies) are used by the mass analyzers to get rid of unwanted ions or enrich desired ions.

There are three major types of mass analyzers that are commonly used in modern instruments: the quadrupole (Q), the ion trap (IT), and the time of flight (TOF) mass analyzers. In the most common configuration for protein analysis, multiple analyzers are combined, generating what is referred to as a tandem mass spectrometer. Two or three analyzers are typically combined in series originating several different configurations.

Likely, the easiest mass analyzer to conceptualize is the time of flight mass spectrometer (reviewed in [12]). In a TOF instrument, ions are separated according to the time they take to travel while accelerated by a magnetic field. The ions hitting the detector are recorded, and this information is presented in a mass spectrum, with mass (m/z—defined below) on the x-axis and the intensity of the signal on the y-axis. To visualize how a TOF mass analyzer works, imagine a bowling ball and a marble sitting side by side in a lane at a bowling alley. If the exact same amount of force is applied to the bowling ball and the marble at precisely the same time, the marble will reach the end of the lane sooner than the bowling ball. Since we can measure the time it takes for this to occur and we know the amount of force that has been applied, we can calculate the mass of the marble and the bowling ball. While the equations look a bit different, for an ion in a TOF instrument, the time of flight is directly proportional to mass. The only conceptual trick is that since we cannot physically push the ions but instead need to use voltage to apply force, the ions will receive energy in a dose equal to the number of charges that they have on the molecule. For example, a molecule with one charge will receive the equivalent of one push of equal energy, whereas a molecule with two charges will receive two pushes of equal energy, and so on. In order to calculate the mass of an unknown peptide, knowing the time (measured) and the force applied, but not charge state (number of charges), other inferences need to be made. Despite the name mass spectrometry, the mass on a mass spectrum is in reality the "m/z," or mass over charge ratio. Indeed a peptide ion flies at a speed, which is in direct proportion to its charge in the instrument. So mass (m) is actually equal to $m+H/z$, where M=mass, H=mass of a proton, and z is the charge. Fortunately with high-resolution mass spectrometers, z (charge state) can be calculated by using the information stored within the isotopic envelope. This is generated by the natural distribution of isotopes and their relative abundance within a peptide chain. For instance, the natural abundance of ^{13}C generates different isotopic forms of the same peptide. The isotopic envelope, which can be observed before correcting for isotopic distribution, is a representation of the natural occurrence of heavier isotope (e.g., ^{13}C). Since peptides are mostly comprised of carbon, hydrogen, nitrogen, and oxygen, we can use the natural abundance of heavy isotopes to determine what the charge state is by looking at what the mass difference is between the light and heavy isotopes (for reference see [13]). For example, the natural mass of carbon is 12.00 Da exactly. For carbon 13, the mass is 13.00 Da. Therefore for a population of ions in a typical peptide, most will be made up of ^{12}C; however, some will have ^{13}C. Therefore, when these different forms are resolved in a mass analyzer, we can see the population with the ^{12}C form and the population with the ^{13}C form. To calculate the charge, we look at the mass difference between the two m/z forms of the population. If the mass difference between one isotope and the next is 1, then there is only a +1 charge, if it is 0.5, then there is a +2 charge and so on. Typically, with electrospray ionization instruments, the charge state is +2 and above. Therefore, an important caveat is that we need an instrument of sufficiently high resolution to resolve the differences between the nearest peaks.

After the TOF, the next mass analyzer that is easiest to conceptualize is a quadrupole mass analyzer. The quadrupole, Q or quad for short, is in essence composed of two couples of parallel rods (four poles) aligned with an axis and equally spaced by 90° angles. If one was to look at them standing on a watch dial, one would be at 12 o'clock, one at 3, one at 6, and one at 9. A radio frequency is applied to the rods, and a current is then applied on top of this. In lay terms, one set of forces is used to nudge ions off the axis, and the other to nudge ions on the axis. For example, if filtering higher masses is desired, just enough energy is applied to keep the mass of interest between the rods—thus lower mass ions will crash into the rods or leave the ion beam, because, like in a TOF instrument, a lighter ion will travel farther with the same force. For higher mass filtering, just enough energy is applied to steer lower masses in the center of the beam, and higher masses will not be moved toward the central axis and will eventually exit the ion beam. The small ions will ping-pong back and forth, but the large ions with initial kinetic energy won't be overcome by the small forces. Thus, by working together, the poles in the quadrupole can act as a mass filter for the masses of interest. To generate a mass spectrum, a quadrupole mass spectrometer has to allow each individual ion to pass through to separate the masses. Since ions are nudged along, the resolution of these instruments tends to be much lower than other instruments and is often used in combination with other mass analyzers in hybrid instruments. One of the most powerful applications of a quadrupole instrument is when three quadrupoles are placed in series, also known as a triple quadrupole (Q_3) instruments. In this case, a particular ion can be selected, the second quadrupole can be used to fragment the ions, and the third quadrupole used to transmit only the resulting fragment ions (also known as product ions or transition ions) to the detector. In this approach, termed selective or multiple reaction monitoring, highly specific "transition" ions can be monitored with incredible gains in signal-to-noise ratios. This is because peptides that have the same mass by chance and are co-eluting (isobaric ions) are eliminated prior to reaching the detector. Selective reaction monitoring is described in detail in a recent review by Gianazza and colleagues [14].

The next major type of mass analyzer is the ion trap mass analyzer [15], which is an evolution beyond a traditional ion trap mass analyzer. A traditional ion trap mass analyzer uses similar principles to a quadrupole, except instead of letting ions pass through the gate or not, a trap keeps the ions trapped in an orbit. To measure a mass spectrum, ions are scanned out of the trap (using the same forces as a quadrupole) to the detector. Alternatively, individual ions can be kept in the trap and all the other ions ejected. An Orbitrap instrument uses some of the same principles as an ion trap, except instead of ions traveling inside of the trap, the ions spin between an outer electrode shell and an inner central axis electrode or spindle. An outer "trap" is usually necessary to load ions from the source region into the Orbitrap to overcome the field generated between the outer shell and inner spindle. Similar to a TOF, the heavier the ion, the farther away it "orbits" the electrode, and the lighter an ion is, the closer the orbit.

The Detector

In order to actually detect ions that have been separated by a mass analyzer, a detector is needed. Now working in reverse order, in the case of the Orbitrap, the detector is built into the trap on opposite sides. This configuration is necessary, since ions moving within a magnetic field generate currents on the outer shell electrode. These signals are picked up on either side of the field, and the signals can be deconvoluted using Fourier transformation to generate exquisitely high-resolution mass spectrum. This high resolution is achieved since the current itself is deconvoluted from the actual path of the ions versus being interpreted from electronics as signals are detected in a TOF instrument. In a quadrupole or a TOF instrument, once ions have been separated and sorted, the signal must be converted from ions to electrons. While detectors can vary in their construction, in the case of non-Orbitrap detectors, ions are sent colliding into charged surfaces that amplify the signal into electrons, photons, or both. The intensity of the signal is then recorded and reported along an axis that is mass to charge or m/z. Knowing that detectors, like any electronic equipment, only work within certain ranges, typically four orders of magnitude or less. This is an important consideration, since the detectors often cannot detect weak signals in the presence of strong signals, and if detectors become saturated with too much signal, they can take some time to "reset." From an experimental standpoint, this means that if the signal is too low, it will not stand out from the electronic noise, and if a signal is too high, you will lose the ability to quantitate if the detector is saturated.

Tandem Mass Spectrometry

The final concept that must be introduced to the reader is tandem mass spectrometry. As mentioned, a tandem mass spectrometer is an instrument where mass spectrometry can be performed in tandem. For most MS applications, a hybrid mass spectrometer is used. For proteomic applications, a modest resolution is required (~10–15,000 resolution) to determine the charge state of multiply charged peptides. Given this resolution requirement, most hybrid mass spectrometers use at least a quadrupole as an analyzer. Only one vendor, Thermo Scientific, owns the patent on the Orbitrap mass analyzer. Generally, the quadrupole is used as a mass analyzer to rapidly select ions for fragmentation, followed by different analyzers (such as TOFs and traps). As mentioned, several different configurations exist on the marketplace including trap-TOFs and other magnetic sector detectors which are beyond the scope of this book.

The most common use of a tandem MS instrument is to first measure the mass and intensity of the analytes (MS) and then to isolate one molecular ion in particular, fragment it, and measure the mass of the fragments (a second MS spectrum). We term this operation MS/MS, MS^2, or tandem MS. Conceptually, there are two types of tandem MS instruments: those that operate in tandem separated by space and those that operate in tandem separated by time. Tandem-in-space instruments carry out the isolation of ions, fragmentation of ions, and measurement of fragment ions in different spaces in the instrument. The Q-TOF is the best example of a tandem-in-space instrument, as the first MS experiment allows all ions to pass through the quadrupole and collision cell and be separated by the TOF. In the second MS experiment, the quadrupole isolates the mass of interest and the ion is fragmented (either in a collision cell or by increasing energy of the ions), and the fragments are separated in the TOF. As electronic components improve, at the time of writing, Q-TOFs can easily operate in the 50–200 Hz range, performing many MS/MS experiments in a second.

The second type of instrument is a tandem-in-time hybrid instrument. In a tandem-in-time experiment, the operations are performed in the same region of the mass spectrometer but at different times. An example would be an ion trap instrument, where ions are first collected and scanned out to perform the MS experiment, and then all ions but the ion of interest scanned out, the ion fragmented and the fragments scanned out for the second MS experiment. Hybrid trap instruments now exist in where ions can be measured in the Orbitrap for the first experiment, and a quadrupole used to collect ions, then the ions are fragmented in the loading trap and fragments measured in the Orbitrap. In this manner, the speed of the instrument operations can be increased significantly, with Orbitrap instruments operating in the 18–20 Hz range. While "slower" than a Q-TOF instrument, the ability of optimizing MS/MS by varying fill times of ions in the trap and the ability to perform additional experiments makes a trap instrument more versatile.

Mass Spectrometry in the Context of HIV-1 Proteomics

Chromatography Considerations

Having covered the principles of mass spectrometry in the preceding section, we can appreciate that tandem mass spectrometry will be the most important application of mass spectrometry for most researchers engaging in HIV-1 proteomic studies. In previous primers on proteomics, we have spent considerable time extolling the virtues of performing extensive protein separation techniques to increase the coverage of proteins [16]. In the context of HIV-1 proteomics, the limited amount of sample available for the investigator precludes the use of protein separation methods given the considerable losses that can occur in most gel-based or chromatography approaches. Reiterating, the sample preparation approaches described above, an in-solution digest with trypsin followed by peptide quantitation is the method of choice for HIV-1 proteomics. Fortunately, we are well beyond the days of slow instrumentation, where often only one or two MS/MS events could be performed per second. With instruments now exceeding 50 Hz, the number of MS/MS events that can be obtained per second reduces the need for extensive multiple dimension protein and peptide fractionation approaches.

Given that most experiments will involve a complex mixture of peptides but be limited to under 10 μg of peptide, the best investment for discovery proteomics uses nano-HPLC methods with long gradients and long columns for separation of peptides. Recently, Hsieh and colleagues published a very elegant study examining the relationships between column and gradient lengths on MS and MS/MS performance showing the performance gains of longer nano-HPLC columns [17]. Indeed, some companies are now marketing 1 m-long nano-HPLC monolithic columns (Dionex) that have exceptional performance. HPLC "chip" systems, which reduce the number of connections and reduce the "dead volume" of connections, are also becoming more and more robust. These systems include offerings from Eskigent/ABSciex (ChiPLC) and Agilent (ChipCube System). The chip systems offer less user variability, as do purchased columns; however, they also tend to be much more expensive. It is at this point though that the investment in off-line desalting and accurate peptide quantitation will protect the investment no matter what choices are made. At minimum most facilities should be able to offer a 30-cm column to perform nano-HPLC separations on. Prior to performing extensive experiments with biological samples, testing the system configuration for performance is a good investment prior to running an extensive experiment. Often, a single sample run in triplicate can help to determine the optimal load for the column and optimal chromatography gradients for the sample. In our laboratory, we routinely profile ~1,500 proteins from 10 μg of peptide from HIV-1 virions and ~3,000 proteins from HIV-1-infected MDMs using a 30-cm 150-μM ID column packed with 3-μM C18 resin with a 300 Å pore size, over a 90-min gradient at 500 nL per min on an ABSciex 5600 instrument (manuscript in preparation). This generic method, with direct loading onto the analytical column, is highly reproducible. Given the limited amount of sample available in a typical experiment involving virions or infected primary cells, nano-HPLC is the method of choice; however, as sources on UHPLC systems continue to improve, the gap between micro-flow and nanoflow HPLC will likely narrow. At the time of writing, approximately tenfold more material must be used with UHPLC to obtain the same limits of detection as nanoflow chromatography methods; however, the increased performance and stability of the UHPLC system warrants consideration.

Mass Spectrometry Acquisition: Qualitative Versus Quantitative Methods

Data-Dependent Analysis

By far the most common type of qualitative mass spectrometry experiment is data-dependent analysis (DDA). In this type of experiment, a full scan of all of the masses is first taken by the instrument (MS), then a specific mass is isolated and fragmented (typically by collision-induced dissociation or CID), and the fragment masses measured (MS/MS). Since peptide separation is occurring in real time, the width of a typical peak is only a few seconds. Now that the mass spectrometry field has moved away from slower instruments operating between 2 and 5 Hz and capable of only performing 1–5 MS/MS events per second, the need for extensive peptide fractionation is lessened. New high-resolution/-performance mass spectrometers now operate at speeds of up to 200 Hz at the time of writing, allowing for the acquisition of much more data in a short period of time. This reduces the probability of missing a peptide stochastically. Given the limited amount of material for an HIV-1 experiment, performing an experiment on an instrument slower than 50 Hz (for a Q-TOF) is simply not recommended. If not practical, then strategies must be considered to either separate proteins prior to digestion or peptides after digestion using different fractionation strategies (reviewed in [18]).

From the reader's perspective, the fundamental goal of a DDA experiment is the acquisition of MS/MS data on as many peptides generated from proteins as possible. DDA experiments are typically semiquantitative. As the speed of acquisition and sampling of peptides increases, the number of times a spectrum is observed can be used as an estimate of the protein abundance. This method, termed spectral counting, is a good start at estimating protein abundance. If biological replicates are available, then this approach can be used along with simple statistical tests between groups to identify proteins that are changing under different conditions.

Label-Free Quantitative Approaches: Spectral Counting and Data-Independent Analysis

We are quickly advancing toward observing more and more of the proteome in each experiment, and the issue of quantitation is often becoming more important than detection of unknown proteins. Since we already know all of the viral proteins in HIV-1, for example, should we bother trying to isolate and identify all of them? Perhaps not. If we have already generated a large database of proteins using traditional (DDA approaches), then we can construct in silico databases based upon the time that a peptide has eluted along with the fragmentation spectrum. Once this database is constructed, then we can perform an experiment where we simply skip to the fragmentation step. Generically, in this type of approach, the instrument measures all of the precursor ion masses (and intensities) and then quickly isolates a range of masses and simultaneously fragments them and measures the fragment ions all together. By mapping the precursor ions and fragment ions back to databases constructed before and not trying to isolate a single ion, we can enhance the sensitivity of detection by approximately tenfold. This type of approach is marketed by different vendors (MSE by Waters, All Ions by Agilent and SWATH by ABSciex to name a few); however, in essence it is taking advantage of higher collision energies and looking for fragment ions that are unique to the peptide of interest. The downside of these methods is that care must be taken to ensure that peptides are normalized properly prior to acquisition on the instrument and that the samples are ran on the same column to ensure that retention times of peptides do not drift. If possible, it is best to run each sample twice, once in DDA mode and then once in data-independent analysis (DIA) mode. In this manner, one obtains the best of both worlds: the accurate quantitation and the ability to identify unknowns in each sample. Another drawback of this method is that at the time of writing, the informatic tools to manage proteomic data generated in this manner are limited and often require investing in the vendor's proprietary software platforms or the installation of open-source software, like open Sequential Windowed Acquisition of All Theoretical Fragment Ion Mass Spectra (SWATH-MS), which can be beyond the capability of most users. Other iterations of these methods exist including what Thermo terms parallel reaction monitoring, and all of these methods have significant advantages over DDA methods. Additionally, these approaches do not suffer from the limitations of labeling chemistries described briefly below. SWATH approaches have already shown utility in the study of HIV-1-infected macrophages [19, 20].

Labeling Approaches

Isobaric Tagging

Prior to DIA, isobaric tags for relative and absolute quantitation (iTRAQ) and similar methods were used to label peptides from different conditions and mix them together during separation. The principle of isobaric mass tags is that they are intact and the same mass during the peptide-labeling step. Once the tagged peptides fragment, fragment masses that are unique to each tag are detectible, and an uncharged "balance" region is liberated along with the peptide fragments. In this way, mixtures up to eight components can be mixed together, an approach referred to as "multiplexing," and quantitated relative to one another in a single experiment with iTRAQ reagents (ABSciex). While we and others in the field have experience with these methods in the context of HIV ([2, 21, 22]), we are now using this method less frequently due to challenges with variability of sample labeling, normalization, and bioinformatic challenges for quantitation. An attractive alternative to iTRAQ reagents is the use of tandem mass tags (TMT) from Pierce. These tags are also isobaric like iTRAQ reagents but come in a number of different covalent chemistries that are available for their use including amine-, cysteine-, and carbonyl-reactive chemistries. TMTs, like iTRAQ reagents, have also been used to study HIV-related neurological disease in synaptosomes [23]. Specialized algorithms are required at the data analysis step to ensure that samples are normalized properly, and careful consideration must be given to the reproducibility of the chemistry so that labeling is consistent between samples. The quality of the reagents also deserves consideration to avoid any degradation of the reporters. Many of the challenges associated with the use of iTRAQ reagents were addressed by Luo and Zhao from a statistical viewpoint [24]. Given the expense of the reagents and challenges with labeling and quantitation, many groups, including ours, are moving toward label-free quantitation as described above.

Stably Incorporated Labeled Amino Acids

One method that merits special mention is the use of heavy amino acids for experiments involving in vitro cultures. The stably incorporated labeled amino acids (SILAC) method is especially powerful for in vitro experiments where a cell can replicate at least seven generations to ensure uniform uptake of the label. This is accomplished by growing cell lines in a tissue culture media that contain a heavy amino acid. This allows for the mixing of peptides from different biological samples in the same MS run. The ratios of proteins can then be determined by comparing the precursor intensities of the "light" to the "heavy" peptide. A nice example of the successful application of this technique was recently published by Barrero and colleagues to examine metabolic pathways altered by HIV-1 viral protein R (Vpr) [25]. While powerful, the major drawback of this technique is that sufficient label must be incorporated to resolve the light and heavy peaks, especially for higher-charged peptides (reviewed in [26]). This can be accomplished by using LysC as a protease; however, this also results in larger peptide fragments that can be difficult to sequence. Another caveat is that cells have to be adapted to serum-free culture conditions, so this may impact results. The same caveats exist as described in the introductory chapter insofar as culturing of HIV-1 and changes in host-protein composition in the progeny virions.

Informatics

Intelligent informatic approaches are essential when dealing with HIV-1. We have therefore dedicated our final chapter to HIV-1 informatics, where we will discuss the aspects in detail (Chap. 6). If the reader has followed the advice outlined in this book, then after making excellent informed choices about sample preparation, chromatography approaches, and instrumentation, they will now have reams of MS/MS data on peptides that need to be identified. The first rule of databases is that if the information is not present in a database, then it will not be found. As for HIV-1, especially in the study of polymorphisms, we address this limitation in the subsequent chapter on HIV-1 sequencing (Chap. 4) along with strategies to built appropriate databases in our HIV-1 informatics chapter (Chap. 6). For example, in our group, we have generated custom databases that contain only the entries for human taxonomy and HIV sequences. A comprehensive strategy is elegantly outlined in subsequent chapters. In the case where careful quantitative information is sought for different mutations, then detailed sequence information must be generated de novo. This point is so important that we discuss it redundantly in this chapter, since many individuals will likely elect to have core facilities execute the proteomic portions of their studies and may skip subsequent chapters. While most core facilities have reasonable search approaches vetted by the reviewers of manuscripts that have been produced using primary data from the facility, many facilities will not be aware of the nuances of data analysis for HIV-1. Two suggestions for the reader are to first ensure that an appropriate database is constructed that will adequately cover viral sequences and second, obtain the search results and load them into either an institutional copy of Scaffold (Proteome Software) or a trial version from the company. For the most part, Scaffold will take the uninformed user to an intermediate level by following standard workflows in the software package. Since the metadata from instrumentation is captured in the search results, the software will harvest these data and will help the user to generate automated reports that are acceptable to the major journals where proteomic research is published. Finally, a special mention needs to be made of the HIV-1 proteomic resources available at BioAfrica [27] (bioafrica.net), which has a comprehensive toolbox for HIV-1 bioinformatics and is an excellent starting point about learning what resources are available for the investigator.

Targeted HIV-1 Proteomics and the Path to Clinical Applications

Selective Reaction Monitoring

In our earlier introduction to mass spectrometry, we introduced the concept of quadrupole mass filters and the triple quadrupole (QQQ) instrument. Conceptually we described a precursor ion being selected in the first quadrupole (Q1), the second quadrupole being used as a collision cell (Q2), and the third quadrupole allowing only the fragment ions specific to the entity of interest to be scanned through the third quadrupole (Q3). The probability of an isobaric (same mass) precursor eluting at the same retention time from the HPLC and having the same product ion is extremely low. Thus, while the overall intensity of the signal is much lower than traditional MS/MS, by using SRM (also known as multiple reaction monitoring or MRM), we can increase the overall signal to noise, so that the limits of detection of most targets can be improved 20–100-fold over traditional methods. Also since these methods are quite adaptable to higher flow rates from HPLC systems that use larger columns and hold more material, often one can use much greater starting material to improve the chance of detecting a target of interest.

Low sample abundance is a recurring theme of this book, and as the guidelines are shifting toward immediate treatment of HIV-1 patients, the chances of obtaining primary virus in great quantity are low. Therefore, SRM approaches provide us with some hope in the field that there may be a place for mass spectrometry in the clinical laboratory helping to inform treatment decisions about HIV-1-infected individuals. While we are years away from this becoming a reality, recently we have used SRM approaches to detect conserved HIV-1 peptides down to the low femtogram level on column. Theoretically, if validated, assays like this could replace expensive amplification-based assays in the clinical laboratory to determine HIV-1 viral load. As we understand more about the sequences leading to HIV-1 drug resistance, in addition to determining viral load, the possibility also exists to look for polymorphisms in viral proteins that are associated with drug resistance.

It is our strong opinion that 40 years after the development of the ELISA, we will start to see the replacement of the immunoassay with affinity-based mass spectrometry methods [28, 29]. As costs decrease for mass spectrometry and the sensitivity is increased, it is not unrealistic for affinity enrichment methods to be used with mass spectrometry detection. This is particularly true of technologies like SISCAPA, which stands for stable isotope standard capture with anti-peptide antibodies, termed by Leigh Anderson, who patented the approach. Briefly, this approach uses antibodies targeting peptides generated after proteolytic cleavage along with heavy synthetic peptides used as a standard for quantitation. Much like a competitive ELISA, displacement of the heavy form of the peptide with the light form provides quantitative information on the analyte. Logical extension of the art allows for many combinations of this fundamental assay including post-capture addition of standard or capture of native proteins with their subsequent digestion. Regardless, these types of approaches allow for the development of MS assays that could examine various post-translational modifications of viral proteins or even allow for the quantitation of host proteins after pulldown and separation of virus particles from the blood.

Quick Start Guide for SRM

SRM assay development and assays that approximate SRM, like parallel reaction monitoring and DIA, described above, are becoming very common. Typically a minimum of two different peptides are used to build a targeted assay for a specific protein. These should be peptides that are unique to the protein of interest. The dominant product ion is typically used for quantitation with the addition of at least one or two qualifying ions (also present in the transition) to ensure that the relative ratios of the ions are consistent, thus reducing the chance of accidentally quantitating an isobaric species that co-eluted. For accurate quantitation of a target, heavy peptides are synthesized commercially that are shifted at least 8–12 Da and spiked into the sample at a known concentration. This mass shift is essential so that the isotopic envelope of the heavy standard doesn't overlap with the native isotopic envelope at higher charge states. The transitions for the heavy peptides are also included along with the transitions for the natural isoforms. Comparing the relative intensity of the heavy internal standard to the measured intensity of the target allows for quantitation. An external standard is used to ensure that the measurements are within the linear range of the detector. Like ELISA's or any other quantitative assay, dilutions may be required to get a target into linear range for quantitation. Due to their specificity, once developed, an SRM assay can be very fast (<5 min) and very inexpensive.

Thinking Back to Our Experiments and Motivations

If executed properly, proteomics now becomes a very powerful tool for the HIV-1 virologist. Concurrent to the time of writing, we have published the first special issue of proteomics on the subject "Virology meets Proteomics" (Proteomics Vol.

15 (2015) No. 12). To our knowledge, this represents one of the first collective works on viral proteomics and includes two publications on HIV-1 proteomics.

In particular, one of the most undiscovered elements of HIV-1 proteomics is studying viral proteins and their posttranslational modifications. In work we published in the early 2000s using two-dimensional gel electrophoresis, we observed several isoelectric shifts of HIV proteins, compatible with phosphorylated forms. Also, the issue of differential cleavage products of protease is yet to be explored. What about pathogenic versus nonpathogenic viruses? Many have shown the essential role of host restriction factors in making virions noninfectious, and others have shown the role of host proteins in making the virus more infectious. Recently, we published a study examining HIV-1 acylation [30], which showed changes in cellular acylation that were impacted by HIV-1 infection. The experimental possibilities are endless. Through careful quantitation and simple mass spectrometry-based experiments, a typical researcher should be well empowered to produce reasonable amounts of materials and biological replicates to use statistics to quantitate differences in their targets of interest. The power of this method is so great that in one experiment, we are now typically observing >1,500 host proteins in HIV-1 with as little as 5 μg of total protein using the methodologies described above. It is our hope that after reading the history of HIV-1 proteomics and the practical guidance provided herein and in other chapters, that we can inspire and educate scientists to become successful and contribute to this quickly growing field.

Alternative Approaches

We would be remiss to not call out to elegant studies that fall under the umbrella of proteomics but use other approaches, like pulldowns or protein arrays. These very powerful technologies are more mature in areas outside of virology; however, the LaBaer group has recently shown the power of these methods to studying an array of different antiviral responses to viruses [31].

Affinity Pulldown Approaches

A major contributor in the area of IP/interaction studies has been performed by Ileana Cristea's group who has performed elegant work using targeted pulldown-based strategies for specific proteins looking for host-proteins that interact with viral protein targets [32, 33]. Her group has used reporter constructs with tags so that not only can one pull down and examine interacting proteins with viral proteins but also examine by microscopy where these interactions are occurring. These types of affinity-based approaches have been applied to identify restriction factors involved in the control of HIV replication, like SAMHD1 [34], by using affinity tags on viral proteins. Others have also performed elegant work using viral clones that express affinity tags to pull down interacting proteins after the infection of various cell lines [35].

Antigen Presentation

Don Hunt has pioneered the concept of major histocompatibility major histocompatibility complex (MHC) presentation for cancer [36]. The same techniques that are being used for characterizing MHC-bound peptides can also be applied to HIV-1 to potentially identify novel antigens that could be used as therapeutic vaccine targets. With new methods being developed to simultaneously profile small molecules and peptides, sample preparation requirements are becoming more streamlined and may minimize extensive processing requirements [37]. This is particularly true for HIV-1 where the virion itself contains peptide bound to class I and class II MHCs. While identification of peptides with nonspecific cleavages is a challenging informatics problem, we strongly believe that there is a great utility in this method for defining how viral proteins are processed into antigens for vaccine development [37]. Informatics tools and approaches for this purpose are described in Chap. 6.

Protein Arrays

Beautiful work has been performed by the group of Bill Robinson at Stanford, showing the power of antigen arrays for antibody characterization over 10 years ago in the HIV field [38], and more recently applied to other viruses [31]. By spotting proteins to an array and characterizing their composition by mass spectrometry, this technique opens the door to understanding antibody development to various elements, either host or viral proteins or modified viral proteins. As technology improves to clone out the variable, diversity, and joining region (VDJ) rearrangements of antibodies, this method shows promise in the identification of neutralizing antibodies and targets that could contribute to the development of sterilizing vaccines [39].

Conclusions

While we are still several years away from mass spectrometry being a "black-box" type of instrument where we simply inject our sample and walk away, rapid recent advancements in mass spectrometry data acquisition and bioinformatics have taken much of the pain out of the path to success. The most fundamentally important aspect of HIV-1 proteomics or any proteomic success is in sample preparation and the accurate quantitation of peptides post-desalting. Subsequent chapters expand in much greater detail, strategies geared toward the measurement of different post-translational modifications of HIV and associated proteins as well as the informatics approaches designed to enhance success.

References

1. Ott DE. Purification of HIV-1 virions by subtilisin digestion or CD45 immunoaffinity depletion for biochemical studies. Methods Mol Biol. 2009;485:15–25.
2. Linde ME, Colquhoun DR, Ubaida Mohien C, Kole T, et al. The conserved set of host proteins incorporated into HIV-1 virions suggests a common egress pathway in multiple cell types. J Proteome Res. 2013;12:2045–54.
3. Cornelissen M, Heeregrave EJ, Zorgdrager F, Pollakis G, et al. Generation of representative primary virus isolates from blood plasma after isolation of HIV-1 with CD44 MicroBeads. Arch Virol. 2010;155:2017–22.
4. Angin M, Klarenbeek PL, King M, Sharma SM, et al. Regulatory T cells expanded from HIV-1-infected individuals maintain phenotype. TCR repertoire and suppressive capacity. PLoS One. 2014;9:e86920.
5. Tanca A, Biosa G, Pagnozzi D, Addis MF, Uzzau S. Comparison of detergent-based sample preparation workflows for LTQ-Orbitrap analysis of the Escherichia coli proteome. Proteomics. 2013;13:2597–607.
6. Mackintosh JA, Veal DA, Karuso P. Fluoroprofile, a fluorescence-based assay for rapid and sensitive quantitation of proteins in solution. Proteomics. 2005;5:4673–7.
7. Feist P, Hummon AB. Proteomic challenges: sample preparation techniques for microgram-quantity protein analysis from biological samples. Int J Mol Sci. 2015;16:3537–63.
8. O'Doherty U, Swiggard WJ, Malim MH. Human immunodeficiency virus type 1 spinoculation enhances infection through virus binding. J Virol. 2000;74:10074–80.
9. Marozsan AJ, Fraundorf E, Abraha A, Baird H, et al. Relationships between infectious titer, capsid protein levels, and reverse transcriptase activities of diverse human immunodeficiency virus type 1 isolates. J Virol. 2004;78:11130–41.
10. Antharavally BS. Removal of detergents from proteins and peptides in a spin-column format. In: Coligan JE et al., editors. Current protocols in protein science; 2012, Chapter 6, Unit 6.12
11. Norris JL, Porter NA, Caprioli RM. Mass spectrometry of intracellular and membrane proteins using cleavable detergents. Anal Chem. 2003;75:6642–7.
12. Cotter RJ, Griffith W, Cotter C. Tandem time-of-flight (TOF/TOF) mass spectrometry and the curved-field reflectron. J Chromatogr B. 2007;855:2–13.
13. Godin JP, Fay LB, Hopfgartner G. Liquid chromatography combined with mass spectrometry for 13C isotopic analysis in life science research. Mass Spectrom Rev. 2007;26:751–74.
14. Gianazza E, Tremoli E, Banfi C. The selected reaction monitoring/multiple reaction monitoring-based mass spectrometry approach for the accurate quantitation of proteins: clinical applications in the cardiovascular diseases. Expert Rev Proteomics. 2014;11:771–88.
15. Scigelova M, Makarov A. Advances in bioanalytical LC-MS using the Orbitrap mass analyzer. Bioanalysis. 2009;1:741–54.
16. Graham DR, Elliott ST, Van Eyk JE. Broad-based proteomic strategies: a practical guide to proteomics and functional screening. J Physiol. 2005;563:1–9.
17. Hsieh EJ, Bereman MS, Durand S, Valaskovic GA, MacCoss MJ. Effects of column and gradient lengths on peak capacity and peptide identification in nanoflow LC-MS/MS of complex proteomic samples. J Am Soc Mass Spectrom. 2013;24:148–53.
18. Yin X, Zhang Y, Liu X, Chen C, et al. Systematic comparison between SDS-PAGE/RPLC and high-/low-pH RPLC coupled tandem mass spectrometry strategies in a whole proteome analysis. Analyst. 2015;140:1314–22.
19. Haverland NA, Fox HS, Ciborowski P. Quantitative proteomics by SWATH-MS reveals altered expression of nucleic acid binding and regulatory proteins in HIV-1-infected macrophages. J Proteome Res. 2014;13:2109–19.
20. Arainga M, Guo D, Wiederin J, Ciborowski P, et al. Opposing regulation of endolysosomal pathways by long-acting nanoformulated antiretroviral therapy and HIV-1 in human macrophages. Retrovirology. 2015;12:5.

21. Guo Y, Singleton PA, Rowshan A, Gucek M, et al. Quantitative proteomics analysis of human endothelial cell membrane rafts: evidence of MARCKS and MRP regulation in the sphingosine 1-phosphate-induced barrier enhancement. Mol Cell Proteomics. 2007;6:689–96.
22. Bregnard C, Zamborlini A, Leduc M, Chafey P, et al. Comparative proteomic analysis of HIV-1 particles reveals a role for Ezrin and EHD4 in the Nef-dependent increase of virus infectivity. J Virol. 2013;87:3729–40.
23. Banerjee S, Liao L, Russo R, Nakamura T, et al. Isobaric tagging-based quantification by mass spectrometry of differentially regulated proteins in synaptosomes of HIV/gp120 transgenic mice: implications for HIV-associated neurodegeneration. Exp Neurol. 2012;236:298–306.
24. Luo R, Zhao H. Protein quantitation using iTRAQ: review on the sources of variations and analysis of nonrandom missingness. Stat Interface. 2012;5:99–107.
25. Barrero CA, Datta PK, Sen S, Deshmane S, et al. HIV-1 Vpr modulates macrophage metabolic pathways: a SILAC-based quantitative analysis. PLoS One. 2013;8:e68376.
26. Dittmar G, Selbach M. SILAC for biomarker discovery. Proteomics Clin Appl. 2015;9:301–6.
27. Doherty RS, De Oliveira T, Seebregts C, Danaviah S, et al. BioAfrica's HIV-1 proteomics resource: combining protein data with bioinformatics tools. Retrovirology. 2005;2:18.
28. Razavi M, Frick LE, LaMarr WA, Pope ME, et al. High-throughput SISCAPA quantitation of peptides from human plasma digests by ultrafast, liquid chromatography-free mass spectrometry. J Proteome Res. 2012;11:5642–9.
29. Anderson NL, Jackson A, Smith D, Hardie D, et al. SISCAPA peptide enrichment on magnetic beads using an in-line bead trap device. Mol Cell Proteomics. 2009;8:995–1005.
30. Colquhoun DR, Lyashkov AE, Mohien CU, Aquino VN, et al. Bioorthogonal mimetics of palmitoyl-CoA and myristoyl-CoA and their subsequent isolation by click chemistry and characterization by mass spectrometry reveal novel acylated host-proteins modified by HIV-1 infection. Proteomics. 2015;15(12):2066–77.
31. Bian X, Wiktor P, Kahn P, Brunner A, et al. Antiviral antibody profiling by high-density protein arrays. Proteomics. 2015;15(12):2136–45.
32. Cristea IM, Williams R, Chait BT, Rout MP. Fluorescent proteins as proteomic probes. Mol Cell Proteomics. 2005;4:1933–41.
33. Diner BA, Lum KK, Javitt A, Cristea IM. Interactions of the antiviral factor IFI16 mediate immune signaling and herpes simplex virus-1 immunosuppression. Mol Cell Proteomics. 2015;14(9):2341–56.
34. Hrecka K, Hao C, Gierszewska M, Swanson SK, et al. Vpx relieves inhibition of HIV-1 infection of macrophages mediated by the SAMHD1 protein. Nature. 2011;474:658–61.
35. Jager S, Cimermancic P, Gulbahce N, Johnson JR, et al. Global landscape of HIV-human protein complexes. Nature. 2012;481:365–70.
36. Appella E, Padlan EA, Hunt DF. Analysis of the structure of naturally processed peptides bound by class I and class II major histocompatibility complex molecules. EXS. 1995;73:105–19.
37. Tharakan R, Tao D, Ubaida-Mohien C, Dinglasan RR, Graham DR. Integrated microfluidic chip and online SCX separation allows untargeted nanoscale metabolomic and peptidomic profiling. J Proteome Res. 2015;14:1621–6.
38. Neuman de Vegvar HE, Amara RR, Steinman L, Utz PJ, et al. Microarray profiling of antibody responses against simian-human immunodeficiency virus: postchallenge convergence of reactivities independent of host histocompatibility type and vaccine regimen. J Virol. 2003;77:11125–38.
39. Robinson WH. Sequencing the functional antibody repertoire—diagnostic and therapeutic discovery. Nat Rev Rheumatol. 2015;11:171–82.

Chapter 5
HIV-1 Sequencing

Shelby L. O'Connor

Introduction

There is a great interest in deep sequencing RNA virus populations, including HIV /SIV, influenza, hepatitis, and Ebola. The RNA polymerases that are critical for replication of many RNA viruses have much lower fidelity than DNA polymerases employed during genome replication [1]. This lower fidelity facilitates the accumulation of nucleotide variants into a progeny viral genome during each replication. This can have enormous consequences on immune evasion, zoonotic events, drug resistance, and pathogenesis [2–7]. With such an enormity of hypotheses to test, there has been an increasing interest in deep sequencing virus populations, and numerous methods have been developed alongside this interest. From a proteomics perspective, these accumulated viral polymorphisms can impact the degree of coverage one can obtain from a proteomics experiment. This chapter focuses exclusively on the evolution of deep sequencing HIV and SIV populations as a means to understand the genetic makeup of these virus populations that contributes to their behavior in vivo. With this understanding the reader should be able to learn how to obtain detailed sequence information to build custom databases that will inform their subsequent proteomics experiments.

S.L. O'Connor, Ph.D. (✉)
Department of Pathology and Laboratory Medicine, University of Wisconsin-Madison, 555 Science Drive, Madison, WI 53711, USA
e-mail: slfeinberg@wisc.edu

D.R.M. Graham, D.E. Ott (eds.), *HIV-1 Proteomics*,
DOI 10.1007/978-1-4939-6542-7_5

Reasons Why Understanding HIV/SIV Variation Is Important from a Biological and Proteomics Perspective

HIV/SIV can evade host immune responses. The first data providing evidence that the accumulation of mutations in HIV can lead to escape from CD8 T cells was presented in 1991 [8] and then shown more clearly a few years later [9–11]. This was followed by data in SIV-infected nonhuman primates to demonstrate that immune escape from CTL occurs during acute and chronic infection [12, 13]. Evidence of HIV/SIV escape from antibodies was also shown on several occasions [14–17]. These important findings required the collection of sequence data from virus populations replicating within individuals. This evidence provided an explanation for why host immune responses are ultimately unable to contain replication of HIV in most infected individuals. Even though immune escape has been well documented, there is an ongoing need to assess the evolutionary path of immune escape in the context of different host genetic backgrounds and in the face of other immune modulatory factors. Glycosylation plays an important role in HIV/SIV host immune evasion [18]. Since many mass spectrometry algorithms reject information lacking consensus glycosylation sequences, for these studies, high sequence coverage is necessary to properly interpret mass spectrometry data.

In addition to evasion of natural immune responses, there are concerns over evasion of HIV from vaccine-elicited immune responses. From past HIV vaccine trials, we have learned that the presence of vaccine-elicited immune responses can select for transmission of virus sequences that are less likely to be neutralized by host immune responses [19–21]. Understanding whether vaccine-elicited immune responses can select for transmission of viruses with specific sequences or affect the emergence of immune escape variants after HIV transmission is essential for evaluating vaccine efficacy. While newer algorithms are much better at assigning polymorphisms from mass spectrometry data alone, having extensive information on polymorphisms present in the virus population assists in the proper assignment of spectra to peptides.

HIV viruses can accumulate mutations such that viral proteins become resistant to antiretroviral medications. Antiretroviral treatment is the most commonly used approach to treat people with HIV. It is well established that strict adherence to antiretroviral drug regimens is essential to prevent the emergence of drug-resistant mutations [22, 23]. To prevent widespread circulation of drug-resistant variants, there is ongoing global surveillance of HIV drug-resistant viruses, and there are efforts to implement low-cost sequencing protocols in these countries to expand these programs [24–27]. By improving our tools to track HIV drug-resistant mutations, there will be better global surveillance of transmitted HIV drug-resistant strains so that numerous individuals are not vulnerable to HIV infection without the option for using antiretrovirals. In this case obtaining sequence information about known mutations related to drug resistance might inform an experiment so that viral enzymes are enriched or targeted in the analysis. Also obtaining detailed sequence information could assist in better quantitation of the relative abundance of the mutation using targeted proteomics.

Globally, circulating strains of HIV are diverse and still evolving. There are at least nine genetic subtypes (clades) and numerous circulating recombinant forms (CRFs) of HIV worldwide [28, 29]. These different subtypes can differ by approximately 30 % in *env*, thus making them quite distinct virus groups [30]. Although it has been speculated that HIV transmission varies for the different subtypes, there is no significant in vivo evidence for this [28]. Still, there is evidence that there are differences in disease progression of antiretroviral-naïve individuals infected with different HIV subtypes and CRFs [29, 31–33]. There is no ecological reason why these different subtypes of HIV will remain localized to their current location; the opposite is true, as there have been several introductions of HIV subtypes into new regions throughout the epidemic [34]. Thus, ongoing monitoring of the prevalence of the different subtypes of HIV will help predict how the epidemic may change at an epidemiological level.

In addition to surveillance of HIV in humans, monitoring viruses in wild nonhuman primates will help either prevent the next zoonotic event or track the origins of the zoonotic event that may occur. There have been at least four instances of zoonotic SIV transmission events leading to the groups of HIV circulating globally [35]. Naturally, this raises concerns that there will be future zoonotic transmission events that will introduce another SIV into the human population or, perhaps, a different RNA virus. Tools to monitor viruses circulating in wild nonhuman primates thus provide a view on the underlying world of existing viruses that may have the potential to jump species. These sorts of surveillance studies are needed to prevent future outbreaks. Thus, we posit that sequencing is the backbone of a successful viral proteomics experiment.

The Evolution of HIV/SIV Sequencing Technologies

The description of the first full-length sequence of HIV (called HTLV-III) in 1985 was performed using Maxam-Gilbert and Sanger techniques [36]. This was followed by Sanger sequencing of the first full-length clone of SIV [37]. These revolutionary studies paved the way for our ability to understand the complex nature of HIV/SIV sequence evolution and diversity.

Over time, Sanger sequencing methods improved [38]. The development of slab gel sequencers that took advantage of fluorescently labeled dideoxynucleotides allowed for higher-throughput DNA sequencing. In many cases, plasmids containing virus genes were sequenced, with the assumption that a single clone was derived from a single virus. Sequence lengths of about 700 bp could be obtained, which took longer, but this approach generated information about linked mutations. At its maximum capacity, the ABI 377 sequencers could sequence 96 samples per 10 h run. This translated into the interrogation of 192 viruses in a 24-h period. The subsequent introduction of capillary DNA sequencing further increased throughput. With an ABI 3730 capillary sequencer, it was possible to sequence 48 plasmids per 2 h run. This translated into the interrogation of 576 viruses in a 24-h period. At that time, these were astounding numbers for sequencing numerous HIV/SIV genomes in a single day.

While sequencing plasmids was informative, it was time consuming and expensive. A transition was then made from sequencing plasmids to sequencing bulk PCR products generated from SIV cDNA. With this approach, it was possible to explore the bulk virus population [39]. Bulk sequencing was limited, however, because it was only possible to identify evidence that a specific nucleotide position was accumulating variants in the total virus population. This approach gave a better perspective of the evolving virus population, but it could not be used to quantify the frequency of a given nucleotide variant in a population.

A benefit of Sanger sequencing is the relatively long length of sequencing reads (approximately 400–700 bp), facilitating analyses of strings of nucleotide sequences. This key advantage has kept the use of Sanger sequencing in favor for single genome amplification (SGA) sequencing of HIV/SIV [40]. SGA approaches have been instrumental in defining transmitted/founder viruses that initiate an HIV or SIV infection [41, 42]. Unfortunately, the cost and time required to sequence large virus populations using this technology has made it somewhat obsolete for characterizing total virus population.

In 2005, a sequencing revolution began with the description of pyrosequencing technology [43]. This discovery opened doors to increase the throughput of sequencing HIV/SIV. With this technology came a transition to pyrosequence HIV/SIV. The GS 20 and GS FLX were early instruments typically found at core facilities, and they were used for initial virus pyrosequencing experiments [44–47]. These were expensive and difficult to use, so Roche/454 developed the benchtop GS Jr. sequencer in 2009 that was adopted by many labs. Using this technology, a single piece of PCR-amplified DNA was attached to a bead, clonally amplified, and then pyrosequenced. This approach yielded about 100,000 sequences on a GS Jr. in a single run that took about 24 h to process. Once the GS Jr. became commonplace, it was used routinely for sequencing virus populations for the next few years [48–52]. Each sequence generated by the Roche/454 pyrosequencing technology was thought to represent a single piece of DNA that was part of a PCR product that was derived from an original virus particle. Of course, PCR amplification of the same virus sequence was still expected. Yet, if all viruses were equally amplified, then the distribution of virus sequences would be a reasonable reflection of the relative viruses in the overall population. Overall, this meant that 100,000 viruses could be interrogated in 24 h.

The length of reads that can be obtained on the Roche/454 platform can vary from 400 to 1000 bp. This is dependent upon the sample preparation and the sequencer being used. This length is on par with that obtained by Sanger technology, so linkage and haplotype information can be obtained from any given read. In contrast, a major problem associated with pyrosequencing was a high error rate frequently associated with the difficulties defining the number of specific nucleotides present in homopolymers [47, 53]. This is especially a major concern when sequencing drug resistance mutations in HIV because many drug resistance mutations occur in homopolymer regions. To accurately characterize mutations in HIV, an assortment of data filtering steps and analysis tools were developed to ensure that a mutation in a homopolymer region was authentic [52, 54, 55]. Consequently, this technique is useable, but not ideal, for widespread screening of drug-resistant HIV.

Sequencing by synthesis using the Illumina sequencing platform is the current workhorse for deep sequencing HIV/SIV. Solexa, the original company to develop sequencing-by-synthesis technology, was formed in 1998. It was acquired by Illumina in 2007, and since then, the use of this technology has increased dramatically [56]. When originally implemented, Solexa Genome Analyzers were used for sequencing but were replaced with Illumina HiSeq machines in 2010. Since then, these machines have been available in core facilities, but are expensive to operate. In 2011, Illumina developed the benchtop MiSeq that was more amenable for use in smaller labs. This machine has greatly improved the accessibility for larger numbers of labs to incorporate deep sequencing in their research portfolios [57]. One drawback to this technology is that it splits DNA into a lot of small pieces, and then the small pieces are sequenced. This makes it difficult to generate long contiguous sequences, and it can be difficult to create analysis pipelines to process the data. Currently, the length of reads ranges from 125 to 300 bp. When merging paired end reads, it is possible to sequence a single DNA fragment of about 500 bp. As these lengths have approached the capability of Sanger sequencing, it has made deep sequencing of HIV/SIV substantially more practical. In addition, sequences generated by this technology are not subject to the same homopolymer errors that plague pyrosequencing. Further, a single MiSeq run can sequence ten million pieces of DNA in a run that lasts 3 days. If each sequence came from a single piece of DNA generated in a PCR amplification, then this means that more than three million viruses can be interrogated in a 24-h period using the Illumina MiSeq.

Given the advantages of the Illumina MiSeq (high fidelity, low cost, increasingly longer read length), there has been a movement to focus development efforts to deep sequence HIV/SIV on the Illumina platform. For this reason, the remainder of this chapter will focus on the methodology and analysis approaches used to sequence HIV/SIV on the Illumina MiSeq.

HIV/SIV Deep Sequencing Methodologies

Pre-experimental Questions

Before initiating any experiment designed to deep sequence HIV or SIV, it is necessary to carefully identify the goals for the experiment and the subsequent goals one wishes to obtain with a complementary mass spectrometry experiment. The following section outlines some key questions to address before undertaking the task of deep sequencing.

Question 1: Am I interested in characterizing SNPs across the entire coding sequence of the genome?

With current technologies, it is difficult to sequence the complete coding region of a single virus without cloning it into a plasmid. It is possible, however, to obtain the frequency of each independent nucleotide at every position throughout the HIV/

SIV coding sequence. For this approach, long amplicons are generated and then fragmented into libraries. The pieces in the libraries are tagged and sequenced. The sequenced pieces can then be mapped back to a reference sequence and the frequency of each nucleotide at each position can be measured. Even though this approach provides little information about linkage, it can provide useful information about ongoing evolution of sequence variants in the virus population that may confer resistance to host immune responses or antiretroviral drugs [26, 45, 58].

Question 2: Do I want to determine linked sequences across a small section of the genome?

It is entirely feasible to obtain linkage information about short stretches of virus sequences, as long as the maximum nucleotide distance is realistically within the limitations of the sequencing methodology. For this approach, relatively small amplicons (~300–500 bp) are generated, tagged, and sequenced from both ends of the DNA in individual clusters. The paired sequences from the same cluster are merged, and then each merged piece of DNA is treated as a single DNA sequence. Sequence information is obtained across the entire piece of DNA and for each piece of DNA from the PCR product that was sequenced. The frequency that a certain stretch of nucleotides appears in the total population can then be calculated. Even though this approach is limited to a small region of the genome, it is an effective way to quantify variation within an entire T cell epitope, investigate linked variants within the envelope gene, or track individual virus populations replicating in an animal.

Question 3: Do I know the sequence of my SIV inoculum or the sequence of the virus population soon after HIV infection?

It is important to know the sequences of HIV/SIV that are replicating during acute infection for two reasons. First, it is important to know the sequences of the "baseline" virus population so that sequences present at later time points can be compared back to early virus sequences. Second, it is important to design primers in conserved virus regions that will amplify virus sequences in an unbiased manner. If the sequences of the early virus populations or the SIV inoculum are not known, these can be obtained through unbiased sequencing of viral cDNA. This approach has been used to discover new strains of SIV and other RNA viruses from the plasma of wild nonhuman primates [3, 48, 59].

Question 4: What is the virus titer?

Knowing the virus titer is important for experimental setup. To sequence SIV and HIV, there obviously needs to be detectable virus so that there is material to sequence. When titers are low, concentrating the virus becomes essential. In addition, long stretches of RNA are difficult to amplify by RT-PCR when there is a low amount of starting material, and thus amplifying shorter segments by RT-PCR is more practical. When titers are high (>10^4 copies/ml of plasma), then RT-PCR occurs readily and yields large amounts of viral cDNA for sequencing. Given the need to amplify the viral cDNA so there is product available for sequencing, it is expected that templates will be resampled. Although idealistic, it is reasonable to expect that the PCR products generated by amplification of templates using primers located in conserved regions are a good representation of the starting virus population [48].

It is useful to consider the following hypothetical example when assessing how virus titer will influence the results. Imagine that there are 20 vRNA templates present at a 50/50 ratio of the variant to the wild type. If only 50 % of these templates are amplified by RT-PCR, then amplification of five wild-type and five variant templates are needed to observe a 50/50 ratio in the final data set. If, by chance, this changes by one template, then four wild-type and six variant templates are amplified, so the ratio of wild type to variant is 40/60 in the final data set. In contrast, imagine that there are 2000 vRNA templates present at a 50/50 ratio of the variant to the wild type. If only 50 % of these templates are amplified by RT-PCR, then amplification of 500 wild-type and 500 variant templates are needed to observe a 50/50 ratio of the variant to the wild type. If, by chance, this changes by one template, then 499 wild-type and 501 variant templates are amplified, so the ratio of wild type to variant will still be approximately 50/50 in the final data set. Thus, slight perturbations in template amplification are less apparent when the titers are high. When analyzing low titer samples, more accurate information about the sequence of the virus population can be obtained by sequencing multiple independent samples.

Sequencing Methodology

Once the above questions have been carefully considered, there are some standard HIV/SIV sequencing pipelines that can be easily modified for a given experimental question. In the following sections, three methodologies will be outlined: (a) unbiased whole genome, (b) PCR-amplified whole genome, and (c) PCR-amplified short segments.

a. Unbiased whole genome

   This approach is typically used to sequence an entire viral genome when very little is known about the virus used in the study and there are no primers available to amplify the virus segments. Initially, viral RNA is isolated from a sample. Random hexamers are used to prime the RNA to initiate synthesis of the first strand of cDNA. The RNA is degraded and then the second strand of cDNA synthesis is completed. The double-stranded cDNA segments are fragmented into a library using the Nextera tagmentation kits (Illumina). Library ID tags are added and the fragments are sequenced on an instrument, such as the MiSeq. Analysis of the data set will be described below.

   This unbiased whole genome approach has been used extensively to discover RNA viruses present in wild nonhuman primates [3, 4, 60–63]. We have also used this approach routinely to sequence virus stocks for clients who have little a priori knowledge of the inoculum sequence. In an experiment comparing data obtained by this unbiased approach vs. amplicon-based approaches (below), the data appears to be quite similar [48]. Notably, the unbiased approach requires a high titer, or there will otherwise be insufficient starting material to yield adequate cDNA of the entire viral genome.

b. Amplicon-based whole genome

This approach is typically used to sequence an entire viral genome when there is an extensive amount of information known about the inocula. Primers specific for conserved sites in the virus genome are designed and then used to amplify viral gene segments by RT-PCR. Similar to the double-stranded cDNA created using the unbiased approach, the double-stranded PCR products are fragmented into libraries with the Nextera technology, tags are added, and then sequenced on an instrument, such as the MiSeq.

This biased sequencing approach is advantageous for many reasons. Unlike the unbiased approach, samples with low viral titers can be used as starting material, a critical need when trying to sequence from individuals who have low viral loads attributed to either natural or drug-mediated control. In addition, there is an extensive amount of flexibility inherent to this approach. As long as a PCR product can be generated from nucleic acid starting material, then it can be fragmented into a library and sequenced. This approach has been used to interrogate the frequencies of nucleotides across the viral genome to quantify variation in T cell epitopes [58, 64]. Besides analyses of entire viral genomes, it is possible to amplify a single gene, such as HIV polymerase, to characterize drug-resistant mutations [26].

One limitation to fragmenting DNA into a library is that sequences of variable length are created that have diverse start and stop positions. These inconsistencies do not affect the mapping of reads to a reference sequence, but they limit the information that can be gained about linkage between sites. In sum, both amplicon-based and unbiased whole genome sequencing are flexible and can yield information about variant frequencies across a wide data set, but they are limited in the information that can be obtained about linkage across a specific area of a genome.

c. Amplicon-based sequencing of a small segment of the viral genome

This approach is typically used to sequence a small segment of a genome when the start and end positions are known and the entire string of nucleic acids across each piece of DNA are of interest. Primers specific for the region of interest are used to amplify the virus segment by RT-PCR. Tags are added that are both unique for the specific sample and help initiate sequencing on the instrument being used. These tags can be added using the TruSeq kits (Illumina) with or without PCR. The addition of tags by PCR leaves open the possibility that bias or errors may be incorporated into the DNA with the additional amplification steps. The addition of tags without PCR leaves open the possibility of greater sample loss. Either way, the tagged samples can be pooled and loaded on an instrument, such as an Illumina MiSeq.

This sequencing approach is advantageous for sequencing entire short segments of DNA that are the same length. On the Illumina MiSeq platform, sequences of up to 300 bp from each end of a single DNA molecule can be generated. After trimming and merging of paired sequences, it is entirely feasible to generate high-quality sequence information for an amplicon of about 400–500 bp long. Information about identical sequences or linked variants can be determined. Ultimately, this data covers a relatively small length of sequence, but a great depth of coverage can be obtained for a segment of DNA that is up to 500 bp.

Sequence Analysis

Gathering sequence data has become relatively easy, but analyzing it can be difficult and time consuming. There are many tools available to help investigators analyze sequence data. Many of these are complex and some require a basic knowledge of the command line. Many biologists who are interested in sequencing, however, are not familiar with the sort of bioinformatics needed to transform sequence information into something that is meaningful to an end user. Fortunately, there is software available for non-bioinformaticians to interact with their data. This chapter will emphasize the use of the software product called Geneious (Biomatters, Ltd). This software has a graphical user interface to visualize aligned reads, making sequence analyses more accessible to non-bioinformaticians. The next section will walk you through a general outline of how data that is obtained from an Illumina MiSeq can be explored in Geneious to obtain a set of information that can be handled in a viewable format.

Prior to sequencing samples, information linking each barcode tag to a sample is entered on the instrument. This is an essential step so that sequence information can be deconvoluted after the run. Once the sample information has been entered, then a DNA pool containing the denatured samples is loaded onto a flow cell, and the single-stranded DNA molecules hybridize to specific adapter oligos on the surface of the flow cell. The hybridized DNA molecules are amplified on the flow cell to form a cluster of single-stranded DNA molecules located at a single coordinate. DNA in each cluster is sequenced so that a single consensus forward and a single consensus reverse sequence read are obtained for the DNA located at each cluster [56].

Trimming, Merging, and Mapping FASTQ Reads

After the sequencing run is complete, the instrument deconvolutes the data by barcode. Two FASTQ files are generated for each barcode: one file containing all the forward reads and one file containing all the reverse reads. FASTQ reads are FASTA reads (strings of nucleotide sequences) with quality information attached to each nucleotide. Each FASTQ read is labeled with a header containing information about the machine, the run number, the coordinate position of the cluster, and the direction of the sequence (http://www.illumina.com/documents/products/techspotlights/techspotlight_sequencing.pdf).

FASTQ reads can be imported into Geneious and then quality trimmed. The trimming stringency may depend on your specific experiment. Once trimmed, the coordinate position associated with each read can be used so that the forward and reverse reads from the same cluster can be interrogated as a single unit. Paired reads can then be merged using a tool called FLASH (Fast Length Adjustment of SHort reads) [65]. There is an available plug-in that can be used for this purpose in Geneious (http://www.geneious.com/plugins/flash).

After trimming and merging, the reads can be mapped to a reference sequence, if available. The appropriate reference sequence should be imported into Geneious. Both the reference and the reads are selected and then the "Map to reference" function can be applied. The mapping settings are specific to the experiment being performed, but there is no harm using one of the default settings to explore the data. There is no perfect group of settings for mapping reads to a reference. By mapping the reads, all the sequences will be arranged in the same direction as the virus sequence. Once reads are mapped, then there is typically interest in either quantifying the frequency of individual SNPs or linked segments in the genome. These two methods are outlined below.

Quantify the Frequency of Individual SNPs Across the Virus Genome

Under the "Annotate & Predict" menu in Geneious, there is an option to "Find Variations/SNPs." After selecting that option, parameters are chosen to detect variants in the mapped sequences, relative to the reference. Some parameters should be considered carefully:

a. Depth of coverage: It is important to determine how much coverage at a given position is required to determine if a variant is authentic. Oftentimes, there is low coverage at the ends of the amplicons that are fragmented and then sequenced. Requiring a minimum amount of coverage is important so detection of false positives is minimized.
b. Minimum variant frequency: This is a highly subjective number. Control experiments have determined that 2 % is a reliable threshold to use to consider a variant authentic or not [26]. This threshold, however, will be somewhat dependent on the technical procedures used upstream of the sequencing reactions. Another consideration is the amount of computer memory required to characterize variants. Choosing a threshold that will detect an informative number of variants without being overwhelming is key.
c. *P*-value: This is a number that takes into account the authenticity of the SNP of interest in the context of the surrounding sequences. An alternative to setting a *p*-value threshold is to have Geneious calculate the *p*-value for the characterized variants. With this approach, the user can consider the *p*-value when determining the authenticity of each variant.

Quantify the Frequency of Linked Segments

Once the alignment is made, portions of sequence reads spanning a specific region of interest can be extracted from the alignment. The region in the reference sequence of interest can be highlighted, and then all the sequences spanning that region can be selected, vertically. These sequence portions can be extracted into a new sequence list. When sequences are extracted from an alignment created from a Nextera library, the reads will be of variable length. When sequences are extracted from an amplicon of a single size, the reads will be of a similar length. Either way, after sequences are extracted, they should be filtered by size, and then reads that are the exact length of the sequence of interest should be extracted. For instance, to examine variation in a nine-amino-acid CD8 T cell epitope, reads that are 27 nucleotides in length should be examined. Once reads of the appropriate length have been extracted, then there are two major analysis routes to choose:

a. Assess the frequency of a nucleotide sequence in the data set: This is a simple counting exercise. In Geneious, there is an option under the Edit menu to "Find duplicates." With this tool, Geneious will identify sequences that are perfectly identical. It will report the number of times a given nucleotide sequence is detected in the total nucleotide sequence list.
b. Assess the frequency of an amino acid sequence in the data set: For this analysis, the reads need to be translated. Under the Sequence menu, there is an option to "Translate." A key point is to ensure the reads are translated in the correct reading frame. Once they are translated, then "Find Duplicates" can be used to determine the number of times a given amino acid sequence is present in the total amino acid sequence list.

In both cases, after the duplicates have been identified, the data can be exported as a FASTA file. The FASTA file can be converted to a tab-delimited file in a text editor. At this point, the frequencies of each nucleotide or amino acid sequence in the population can be calculated.

When measuring variant frequencies, it is important to determine the threshold frequency of the variants to consider as authentic. This is a common concern among researchers and is dependent on a given type of experiment. To determine a minimum threshold for variation, it is best to sequence a sample that should be clonal and has been prepared in the same conditions as the experimental samples. In one example, an HIV clonal stock from a plasmid was created and used to quantify the variant threshold when deep sequencing by the Illumina platform. In this study, a variant frequency of 2 % was found to be a conservative threshold for determining the authenticity of a variant [26]. The threshold to use, however, is dependent upon the preparation protocol, such that empirically measuring the error for an experimental technique is useful.

If possible, detecting a variant in multiple samples from the same individual further increases confidence that the variant is authentic. Longitudinal samples are ideal to assess whether a variant present at a low frequency at an early time point will then expand at a later time point. This type of observation increases the likelihood that the initial variant present at 2 % (or less) is authentic.

De Novo Assembly

There are times when information about the reference sequences is unknown, so it is not possible to map reads to a reference. This is common when using the unbiased sequencing approach described above. In these cases, a de novo assembly of the reads may be appropriate. When performing a de novo assembly, reads will be arranged so that overlapping segments will be linked together to form a single contig. Parameters need to be selected for running a de novo assembly. These parameters are likely going to be specific to a given experiment, but an investigator should not be afraid to use default parameters to explore the data set. Geneious will generate several de novo assembled contigs. For each contig, the closest identity of the consensus sequence can be determined using BLAST. There is an option to perform a "Sequence Search" in Geneious. With this tool, the consensus can be used in a BLAST search against a number of databases. The end user can then explore the hits to determine the identity of the sequence contig.

Discussion and Final Thoughts

Throughout this chapter, a variety of points to consider when deep sequencing HIV/SIV have been raised. Simply saying "I want to deep sequence SIV or HIV" is not enough when designing one of these experiments. Different deep sequencing approaches will yield different data sets that will inform subsequent mass spectrometry experiments in a different contextual manner. While collecting the data may be relatively easy, it is useful to consider how data will be analyzed and stored before beginning.

It is also important to consider whether deep sequencing is appropriate for the specific experimental needs. For instance, should single genome amplification (SGA) be used rather than deep sequencing? The differences between these techniques are substantial. For SGA, vRNA is diluted so that a single amplified product will have been generated from a single virus [66]. This amplified product is sequenced with Sanger technologies. While expensive and time consuming, the sequence data obtained can be attributed to a single template, which may be important for the experiment. In contrast, generating an amplicon and then deep sequencing the amplicon will generate a lot of data about SNP frequencies, but information about linkage between distant sites will be lost.

There is a lot of time and expense associated with deep sequencing. A single run on an Illumina instrument is expensive. Multiplexing samples significantly reduces the cost per sample, but, oftentimes, methods need to be tested on a few samples before expanding to a larger cohort. Testing these sequencing methodologies is made easier when there are colleagues with whom samples can be tested on other ongoing runs. Nonetheless, planning for method development is key, since most SIV/HIV deep sequencing experiments require a custom approach.

Analysis of data must also be considered. Each MiSeq run will generate multiple gigabytes of data. After performing an analysis of the data in Geneious, project sizes can be tens of gigabytes in size. The processing time of large data sets gets longer and becomes overwhelming. Managing, storing, and backing up this data with appropriate hardware needs to be considered in any grant budget.

Besides storing and analyzing data, it can be difficult to display the data. Variation at 10,000 nucleotides across a genome is difficult to display on a single screen. Often, key pieces of information are extracted and then put into a graph or a table to present to an audience. To explore the entire genome, there are some tools available. At UW-Madison, a program called LayerCake was developed to explore and compare variation in multiple genomes [63, 67]. Alternately the V-phaser and V-profiler programs that were developed by the Broad Institute can be used to generate heat maps of the data sets [58, 68].

Predicting the future of HIV/SIV deep sequencing is impossible. Major improvements will come as read lengths increase. While it would be ideal to deep sequence entire virus genomes from the beginning of the virus transcript to the end, this is unlikely to happen in the near future. In addition to read length limitations, the amplification of entire viral genomes without recombination is technically difficult. Single virus template sequencing will likely require new technologies that have not yet been developed.

Still, the technologies available to deep sequences SIV/HIV are useful and can be used to sequence other RNA viruses. The exponential improvements in sequencing technologies are continuing to diversify the hypotheses that are being tested with deep sequencing experiments, and thus our understanding of virus populations will continue to evolve. What is unlikely to change in the near future is the need for better sequence databases to maximize the return on investment for mass spectrometry experiments.

References

1. te Velthuis AJ. Common and unique features of viral RNA-dependent polymerases. Cell Mol Life Sci. 2014;71:4403–20.
2. Gire SK, Goba A, Andersen KG, Sealfon RS, Park DJ, Kanneh L, Jalloh S, Momoh M, Fullah M, Dudas G, Wohl S, Moses LM, Yozwiak NL, Winnicki S, Matranga CB, Malboeuf CM, Qu J, Gladden AD, Schaffner SF, Yang X, Jiang PP, Nekoui M, Colubri A, Coomber MR, Fonnie M, Moigboi A, Gbakie M, Kamara FK, Tucker V, Konuwa E, Saffa S, Sellu J, Jalloh AA, Kovoma A, Koninga J, Mustapha I, Kargbo K, Foday M, Yillah M, Kanneh F, Robert W, Massally JL, Chapman SB, Bochicchio J, Murphy C, Nusbaum C, Young S, Birren BW, Grant DS, Scheiffelin JS, Lander ES, Happi C, Gevao SM, Gnirke A, Rambaut A, Garry RF, Khan SH, Sabeti PC. Genomic surveillance elucidates Ebola virus origin and transmission during the 2014 outbreak. Science. 2014;345:1369–72.
3. Lauck M, Switzer WM, Sibley SD, Hyeroba D, Tumukunde A, Weny G, Taylor B, Shankar A, Ting N, Chapman CA, Friedrich TC, Goldberg TL, O'Connor DH. Discovery and full genome characterization of two highly divergent simian immunodeficiency viruses infecting black-and-white colobus monkeys (*Colobus guereza*) in Kibale National Park, Uganda. Retrovirology. 2013;10:107.

4. Lauck M, Switzer WM, Sibley SD, Hyeroba D, Tumukunde A, Weny G, Shankar A, Greene JM, Ericsen AJ, Zheng H, Ting N, Chapman CA, Friedrich TC, Goldberg TL, O'Connor DH. Discovery and full genome characterization of a new SIV lineage infecting red-tailed guenons (*Cercopithecus ascanius schmidti*) in Kibale National Park, Uganda. Retrovirology. 2014;11:55.
5. Wilker PR, Dinis JM, Starrett G, Imai M, Hatta M, Nelson CW, O'Connor DH, Hughes AL, Neumann G, Kawaoka Y, Friedrich TC. Selection on haemagglutinin imposes a bottleneck during mammalian transmission of reassortant H5N1 influenza viruses. Nat Commun. 2013;4:2636.
6. Newman RM, Kuntzen T, Weiner B, Berical A, Charlebois P, Kuiken C, Murphy DG, Simmonds P, Bennett P, Lennon NJ, Birren BW, Zody MC, Allen TM, Henn MR. Whole genome pyrosequencing of rare hepatitis C virus genotypes enhances subtype classification and identification of naturally occurring drug resistance variants. J Infect Dis. 2013;208:17–31.
7. Ram D, Leshkowitz D, Gonzalez D, Forer R, Levy I, Chowers M, Lorber M, Hindiyeh M, Mendelson E, Mor O. Evaluation of GS Junior and MiSeq next-generation sequencing technologies as an alternative to Trugene population sequencing in the clinical HIV laboratory. J Virol Methods. 2015;212:12–6.
8. Phillips RE, Rowland-Jones S, Nixon DF, Gotch FM, Edwards JP, Ogunlesi AO, Elvin JG, Rothbard JA, Bangham CR, Rizza CR, et al. Human immunodeficiency virus genetic variation that can escape cytotoxic T cell recognition. Nature. 1991;354:453–9.
9. Goulder PJ, Phillips RE, Colbert RA, McAdam S, Ogg G, Nowak MA, Giangrande P, Luzzi G, Morgan B, Edwards A, McMichael AJ, Rowland-Jones S. Late escape from an immunodominant cytotoxic T-lymphocyte response associated with progression to AIDS. Nat Med. 1997;3:212–7.
10. Borrow P, Lewicki H, Wei X, Horwitz MS, Peffer N, Meyers H, Nelson JA, Gairin JE, Hahn BH, Oldstone MB, Shaw GM. Antiviral pressure exerted by HIV-1-specific cytotoxic T lymphocytes (CTLs) during primary infection demonstrated by rapid selection of CTL escape virus [see comments]. Nat Med. 1997;3:205–11.
11. Price DA, Goulder PJ, Klenerman P, Sewell AK, Easterbrook PJ, Troop M, Bangham CR, Phillips RE. Positive selection of HIV-1 cytotoxic T lymphocyte escape variants during primary infection. Proc Natl Acad Sci U S A. 1997;94:1890–5.
12. Evans DT, O'Connor DH, Jing P, Dzuris JL, Sidney J, da Silva J, Allen TM, Horton H, Venham JE, Rudersdorf RA, Vogel T, Pauza CD, Bontrop RE, DeMars R, Sette A, Hughes AL, Watkins DI. Virus-specific cytotoxic T-lymphocyte responses select for amino-acid variation in simian immunodeficiency virus Env and Nef. Nat Med. 1999;5:1270–6.
13. Allen TM, O'Connor DH, Jing P, Dzuris JL, Mothe BR, Vogel TU, Dunphy E, Liebl ME, Emerson C, Wilson N, Kunstman KJ, Wang X, Allison DB, Hughes AL, Desrosiers RC, Altman JD, Wolinsky SM, Sette A, Watkins DI. Tat-specific cytotoxic T lymphocytes select for SIV escape variants during resolution of primary viraemia. Nature. 2000;407:386–90.
14. Richman DD, Wrin T, Little SJ, Petropoulos CJ. Rapid evolution of the neutralizing antibody response to HIV type 1 infection. Proc Natl Acad Sci U S A. 2003;100:4144–9.
15. Wei X, Decker JM, Wang S, Hui H, Kappes JC, Wu X, Salazar-Gonzalez JF, Salazar MG, Kilby JM, Saag MS, Komarova NL, Nowak MA, Hahn BH, Kwong PD, Shaw GM. Antibody neutralization and escape by HIV-1. Nature. 2003;422:307–12.
16. Roederer M, Keele BF, Schmidt SD, Mason RD, Welles HC, Fischer W, Labranche C, Foulds KE, Louder MK, Yang ZY, Todd JP, Buzby AP, Mach LV, Shen L, Seaton KE, Ward BM, Bailer RT, Gottardo R, Gu W, Ferrari G, Alam SM, Denny TN, Montefiori DC, Tomaras GD, Korber BT, Nason MC, Seder RA, Koup RA, Letvin NL, Rao SS, Nabel GJ, Mascola JR. Immunological and virological mechanisms of vaccine-mediated protection against SIV and HIV. Nature. 2014;505:502–8.
17. Wu F, Ourmanov I, Kuwata T, Goeken R, Brown CR, Buckler-White A, Iyengar R, Plishka R, Aoki ST, Hirsch VM. Sequential evolution and escape from neutralization of simian immunodeficiency virus SIVsmE660 clones in rhesus macaques. J Virol. 2012;86:8835–47.

18. Vigerust DJ, Shepherd VL. Virus glycosylation: role in virulence and immune interactions. Trends Microbiol. 2007;15:211–8.
19. Edlefsen PT, Rolland M, Hertz T, Tovanabutra S, Gartland AJ, deCamp AC, Magaret CA, Ahmed H, Gottardo R, Juraska M, McCoy C, Larsen BB, Sanders-Buell E, Carrico C, Menis S, Bose M, Arroyo MA, O'Connell RJ, Nitayaphan S, Pitisuttithum P, Kaewkungwal J, Rerks-Ngarm S, Robb ML, Kirys T, Georgiev IS, Kwong PD, Scheffler K, Pond SL, Carlson JM, Michael NL, Schief WR, Mullins JI, Kim JH, Gilbert PB. Comprehensive sieve analysis of breakthrough HIV-1 sequences in the RV144 vaccine efficacy trial. PLoS Comput Biol. 2015;11:e1003973.
20. Rolland M, Tovanabutra S, deCamp AC, Frahm N, Gilbert PB, Sanders-Buell E, Heath L, Magaret CA, Bose M, Bradfield A, O'Sullivan A, Crossler J, Jones T, Nau M, Wong K, Zhao H, Raugi DN, Sorensen S, Stoddard JN, Maust BS, Deng W, Hural J, Dubey S, Michael NL, Shiver J, Corey L, Li F, Self SG, Kim J, Buchbinder S, Casimiro DR, Robertson MN, Duerr A, McElrath MJ, McCutchan FE, Mullins JI. Genetic impact of vaccination on breakthrough HIV-1 sequences from the STEP trial. Nat Med. 2011;17:366–71.
21. Rolland M, Edlefsen PT, Larsen BB, Tovanabutra S, Sanders-Buell E, Hertz T, deCamp AC, Carrico C, Menis S, Magaret CA, Ahmed H, Juraska M, Chen L, Konopa P, Nariya S, Stoddard JN, Wong K, Zhao H, Deng W, Maust BS, Bose M, Howell S, Bates A, Lazzaro M, O'Sullivan A, Lei E, Bradfield A, Ibitamuno G, Assawadarachai V, O'Connell RJ, deSouza MS, Nitayaphan S, Rerks-Ngarm S, Robb ML, McLellan JS, Georgiev I, Kwong PD, Carlson JM, Michael NL, Schief WR, Gilbert PB, Mullins JI, Kim JH. Increased HIV-1 vaccine efficacy against viruses with genetic signatures in Env V2. Nature. 2012;490:417–20.
22. Hosseinipour MC, Gupta RK, Van Zyl G, Eron JJ, Nachega JB. Emergence of HIV drug resistance during first- and second-line antiretroviral therapy in resource-limited settings. J Infect Dis. 2013;207 Suppl 2:S49–56.
23. Gross R, Yip B, Lo Re V, Wood E, Alexander CS, Harrigan PR, Bangsberg DR, Montaner JS, Hogg RS. A simple, dynamic measure of antiretroviral therapy adherence predicts failure to maintain HIV-1 suppression. J Infect Dis. 2006;194:1108–14.
24. Bennett DE, Bertagnolio S, Sutherland D, Gilks CF. The World Health Organization's global strategy for prevention and assessment of HIV drug resistance. Antivir Ther. 2008;13 Suppl 2:1–13.
25. Jordan MR, Bennett DE, Wainberg MA, Havlir D, Hammer S, Yang C, Morris L, Peeters M, Wensing AM, Parkin N, Nachega JB, Phillips A, De Luca A, Geng E, Calmy A, Raizes E, Sandstrom P, Archibald CP, Perriens J, McClure CM, Hong SY, McMahon JH, Dedes N, Sutherland D, Bertagnolio S. Update on World Health Organization HIV drug resistance prevention and assessment strategy: 2004–2011. Clin Infect Dis. 2012;54 Suppl 4:S245–9.
26. Dudley DM, Bailey AL, Mehta SH, Hughes AL, Kirk GD, Westergaard RP, O'Connor DH. Cross-clade simultaneous HIV drug resistance genotyping for reverse transcriptase, protease, and integrase inhibitor mutations by Illumina MiSeq. Retrovirology. 2014;11:122.
27. Estill J, Salazar-Vizcaya L, Blaser N, Egger M, Keiser O. The cost-effectiveness of monitoring strategies for antiretroviral therapy of HIV infected patients in resource-limited settings: software tool. PLoS One. 2015;10:e0119299.
28. Taylor BS, Sobieszczyk ME, McCutchan FE, Hammer SM. The challenge of HIV-1 subtype diversity. N Engl J Med. 2008;358:1590–602.
29. Hemelaar J. The origin and diversity of the HIV-1 pandemic. Trends Mol Med. 2012;18:182–92.
30. Shaw, GM, Hunter, E. 2012. HIV transmission. Cold Spring Harb Perspect Med 2
31. Easterbrook PJ, Smith M, Mullen J, O'Shea S, Chrystie I, de Ruiter A, Tatt ID, Geretti AM, Zuckerman M. Impact of HIV-1 viral subtype on disease progression and response to antiretroviral therapy. J Int AIDS Soc. 2010;13:4.
32. Pant Pai N, Shivkumar S, Cajas JM. Does genetic diversity of HIV-1 non-B subtypes differentially impact disease progression in treatment-naive HIV-1-infected individuals? A systematic review of evidence: 1996–2010. J Acquir Immune Defic Syndr. 2012;59:382–8.

33. Tarosso LF, Sanabani SS, Ribeiro SP, Sauer MM, Tomiyama HI, Sucupira MC, Diaz RS, Sabino EC, Kalil J, Kallas EG. Short communication: HIV type 1 subtype BF leads to faster CD4+ T cell loss compared to subtype B. AIDS Res Hum Retroviruses. 2014;30:190–4.
34. Tebit DM, Arts EJ. Tracking a century of global expansion and evolution of HIV to drive understanding and to combat disease. Lancet Infect Dis. 2011;11:45–56.
35. Hahn BH, Shaw GM, De Cock KM, Sharp PM. AIDS as a zoonosis: scientific and public health implications. Science. 2000;287:607–14.
36. Ratner L, Haseltine W, Patarca R, Livak KJ, Starcich B, Josephs SF, Doran ER, Rafalski JA, Whitehorn EA, Baumeister K, et al. Complete nucleotide sequence of the AIDS virus, HTLV-III. Nature. 1985;313:277–84.
37. Regier DA, Desrosiers RC. The complete nucleotide sequence of a pathogenic molecular clone of simian immunodeficiency virus. AIDS Res Hum Retroviruses. 1990;6:1221–31.
38. Springer M. Applied biosystems: celebrating 25 years of advancing science. Am Lab. 2006;38(11):4–8.
39. O'Connor DH, Allen TM, Vogel TU, Jing P, DeSouza IP, Dodds E, Dunphy EJ, Melsaether C, Mothe B, Yamamoto H, Horton H, Wilson N, Hughes AL, Watkins DI. Acute phase cytotoxic T lymphocyte escape is a hallmark of simian immunodeficiency virus infection. Nat Med. 2002;8:493–9.
40. Keele BF. Identifying and characterizing recently transmitted viruses. Curr Opin HIV AIDS. 2010;5:327–34.
41. Keele BF, Li H, Learn GH, Hraber P, Giorgi EE, Grayson T, Sun C, Chen Y, Yeh WW, Letvin NL, Mascola JR, Nabel GJ, Haynes BF, Bhattacharya T, Perelson AS, Korber BT, Hahn BH, Shaw GM. Low-dose rectal inoculation of rhesus macaques by SIVsmE660 or SIVmac251 recapitulates human mucosal infection by HIV-1. J Exp Med. 2009;206:1117–34.
42. Keele BF, Giorgi EE, Salazar-Gonzalez JF, Decker JM, Pham KT, Salazar MG, Sun C, Grayson T, Wang S, Li H, Wei X, Jiang C, Kirchherr JL, Gao F, Anderson JA, Ping LH, Swanstrom R, Tomaras GD, Blattner WA, Goepfert PA, Kilby JM, Saag MS, Delwart EL, Busch MP, Cohen MS, Montefiori DC, Haynes BF, Gaschen B, Athreya GS, Lee HY, Wood N, Seoighe C, Perelson AS, Bhattacharya T, Korber BT, Hahn BH, Shaw GM. Identification and characterization of transmitted and early founder virus envelopes in primary HIV-1 infection. Proc Natl Acad Sci U S A. 2008;105:7552–7.
43. Margulies M, Egholm M, Altman WE, Attiya S, Bader JS, Bemben LA, Berka J, Braverman MS, Chen YJ, Chen Z, Dewell SB, Du L, Fierro JM, Gomes XV, Godwin BC, He W, Helgesen S, Ho CH, Irzyk GP, Jando SC, Alenquer ML, Jarvie TP, Jirage KB, Kim JB, Knight JR, Lanza JR, Leamon JH, Lefkowitz SM, Lei M, Li J, Lohman KL, Lu H, Makhijani VB, McDade KE, McKenna MP, Myers EW, Nickerson E, Nobile JR, Plant R, Puc BP, Ronan MT, Roth GT, Sarkis GJ, Simons JF, Simpson JW, Srinivasan M, Tartaro KR, Tomasz A, Vogt KA, Volkmer GA, Wang SH, Wang Y, Weiner MP, Yu P, Begley RF, Rothberg JM. Genome sequencing in microfabricated high-density picolitre reactors. Nature. 2005;437:376–80.
44. Bimber BN, Chugh P, Giorgi EE, Kim B, Almudevar AL, Dewhurst S, O'Connor DH, Lee HY. Nef gene evolution from a single transmitted strain in acute SIV infection. Retrovirology. 2009;6:57.
45. Bimber BN, Dudley DM, Lauck M, Becker EA, Chin EN, Lank SM, Grunenwald HL, Caruccio NC, Maffitt M, Wilson NA, Reed JS, Sosman JM, Tarosso LF, Sanabani S, Kallas EG, Hughes AL, O'Connor DH. Whole-genome characterization of human and simian immunodeficiency virus intrahost diversity by ultradeep pyrosequencing. J Virol. 2010;84:12087–92.
46. Cale EM, Hraber P, Giorgi EE, Fischer W, Bhattacharya T, Leitner T, Yeh WW, Gleasner C, Green LD, Han CS, Korber B, Letvin NL. Epitope-specific CD8+ T lymphocytes cross-recognize mutant simian immunodeficiency virus (SIV) sequences but fail to contain very early evolution and eventual fixation of epitope escape mutations during SIV infection. J Virol. 2011;85:3746–57.

47. Rozera G, Abbate I, Bruselles A, Vlassi C, D'Offizi G, Narciso P, Chillemi G, Prosperi M, Ippolito G, Capobianchi MR. Massively parallel pyrosequencing highlights minority variants in the HIV-1 env quasispecies deriving from lymphomonocyte sub-populations. Retrovirology. 2009;6:15.
48. Hughes AL, Becker EA, Lauck M, Karl JA, Braasch AT, O'Connor DH, O'Connor SL. SIV genome-wide pyrosequencing provides a comprehensive and unbiased view of variation within and outside CD8 T lymphocyte epitopes. PLoS One. 2012;7:e47818.
49. O'Connor SL, Becker EA, Weinfurter JT, Chin EN, Budde ML, Gostick E, Correll M, Gleicher M, Hughes AL, Price DA, Friedrich TC, O'Connor DH. Conditional CD8+ T cell escape during acute simian immunodeficiency virus infection. J Virol. 2012;86:605–9.
50. Brumme CJ, Huber KD, Dong W, Poon AF, Harrigan PR, Sluis-Cremer N. Replication fitness of multiple nonnucleoside reverse transcriptase-resistant HIV-1 variants in the presence of etravirine measured by 454 deep sequencing. J Virol. 2013;87:8805–7.
51. Avidor B, Girshengorn S, Matus N, Talio H, Achsanov S, Zeldis I, Fratty IS, Katchman E, Brosh-Nissimov T, Hassin D, Alon D, Bentwich Z, Yust I, Amit S, Forer R, Vulih Shultsman I, Turner D. Evaluation of a benchtop HIV ultradeep pyrosequencing drug resistance assay in the clinical laboratory. J Clin Microbiol. 2013;51:880–6.
52. Dudley DM, Chin EN, Bimber BN, Sanabani SS, Tarosso LF, Costa PR, Sauer MM, Kallas EG, O'Connor DH. Low-cost ultra-wide genotyping using Roche/454 pyrosequencing for surveillance of HIV drug resistance. PLoS One. 2012;7:e36494.
53. Shao W, Boltz VF, Spindler JE, Kearney MF, Maldarelli F, Mellors JW, Stewart C, Volfovsky N, Levitsky A, Stephens RM, Coffin JM. Analysis of 454 sequencing error rate, error sources, and artifact recombination for detection of low-frequency drug resistance mutations in HIV-1 DNA. Retrovirology. 2013;10:18.
54. Iyer S, Bouzek H, Deng W, Larsen B, Casey E, Mullins JI. Quality score based identification and correction of pyrosequencing errors. PLoS One. 2013;8:e73015.
55. Macalalad AR, Zody MC, Charlebois P, Lennon NJ, Newman RM, Malboeuf CM, Ryan EM, Boutwell CL, Power KA, Brackney DE, Pesko KN, Levin JZ, Ebel GD, Allen TM, Birren BW, Henn MR. Highly sensitive and specific detection of rare variants in mixed viral populations from massively parallel sequence data. PLoS Comput Biol. 2012;8:e1002417.
56. Balasubramanian S. Sequencing nucleic acids: from chemistry to medicine. Chem Commun (Camb). 2011;47:7281–6.
57. Liu L, Li Y, Li S, Hu N, He Y, Pong R, Lin D, Lu L, Law M. Comparison of next-generation sequencing systems. J Biomed Biotechnol. 2012;2012:251364.
58. Adnan S, Colantonio AD, Yu Y, Gillis J, Wong FE, Becker EA, Piatak MJ, Reeves RK, Lifson JD, O'Connor SL, Johnson RP. CD8 T Cell Response Maturation Defined by Anentropic Specificity and Repertoire Depth Correlates with SIVDeltanef-induced Protection. PLoS Pathog. 2015;11:e1004633.
59. Lauck M, Sibley SD, Hyeroba D, Tumukunde A, Weny G, Chapman CA, Ting N, Switzer WM, Kuhn JH, Friedrich TC, O'Connor DH, Goldberg TL. Exceptional simian hemorrhagic fever virus diversity in a wild African primate community. J Virol. 2013;87:688–91.
60. Sibley SD, Lauck M, Bailey AL, Hyeroba D, Tumukunde A, Weny G, Chapman CA, O'Connor DH, Goldberg TL, Friedrich TC. Discovery and characterization of distinct simian pegiviruses in three wild African Old World monkey species. PLoS One. 2014;9:e98569.
61. Goldberg TL, Gendron-Fitzpatrick A, Deering KM, Wallace RS, Clyde VL, Lauck M, Rosen GE, Bennett AJ, Greiner EC, O'Connor DH. Fatal metacestode infection in Bornean orangutan caused by unknown Versteria species. Emerg Infect Dis. 2014;20:109–13.
62. Lauck M, Palacios G, Wiley MR, Li Y, Fang Y, Lackemeyer MG, Cai Y, Bailey AL, Postnikova E, Radoshitzky SR, Johnson RF, Alkhovsky SV, Deriabin PG, Friedrich TC, Goldberg TL, Jahrling PB, O'Connor DH, Kuhn JH. Genome sequences of simian hemorrhagic fever virus variant NIH LVR42-0/M6941 isolates (Arteriviridae: Arterivirus). Genome Announc. 2014;2(5):e00978–14.

63. Bailey AL, Lauck M, Weiler A, Sibley SD, Dinis JM, Bergman Z, Nelson CW, Correll M, Gleicher M, Hyeroba D, Tumukunde A, Weny G, Chapman C, Kuhn JH, Hughes AL, Friedrich TC, Goldberg TL, O'Connor DH. High genetic diversity and adaptive potential of two simian hemorrhagic fever viruses in a wild primate population. PLoS One. 2014;9:e90714.
64. Harris M, Burns CM, Becker EA, Braasch AT, Gostick E, Johnson RC, Broman KW, Price DA, Friedrich TC, O'Connor SL. Acute-phase CD8 T cell responses that select for escape variants are needed to control live attenuated simian immunodeficiency virus. J Virol. 2013;87:9353–64.
65. Magoc T, Salzberg SL. FLASH: fast length adjustment of short reads to improve genome assemblies. Bioinformatics. 2011;27:2957–63.
66. Salazar-Gonzalez JF, Bailes E, Pham KT, Salazar MG, Guffey MB, Keele BF, Derdeyn CA, Farmer P, Hunter E, Allen S, Manigart O, Mulenga J, Anderson JA, Swanstrom R, Haynes BF, Athreya GS, Korber BT, Sharp PM, Shaw GM, Hahn BH. Deciphering human immunodeficiency virus type 1 transmission and early envelope diversification by single-genome amplification and sequencing. J Virol. 2008;82:3952–70.
67. Correll M, Ghosh S, O'Connor D, Gleicher M. Visualizing virus population variability from next generation sequencing data. In: Proceedings of BioVis; 2011
68. Henn MR, Boutwell CL, Charlebois P, Lennon NJ, Power KA, Macalalad AR, Berlin AM, Malboeuf CM, Ryan EM, Gnerre S, Zody MC, Erlich RL, Green LM, Berical A, Wang Y, Casali M, Streeck H, Bloom AK, Dudek T, Tully D, Newman R, Axten KL, Gladden AD, Battis L, Kemper M, Zeng Q, Shea TP, Gujja S, Zedlack C, Gasser O, Brander C, Hess C, Gunthard HF, Brumme ZL, Brumme CJ, Bazner S, Rychert J, Tinsley JP, Mayer KH, Rosenberg E, Pereyra F, Levin JZ, Young SK, Jessen H, Altfeld M, Birren BW, Walker BD, Allen TM. Whole genome deep sequencing of HIV-1 reveals the impact of early minor variants upon immune recognition during acute infection. PLoS Pathog. 2012;8:e1002529.

Chapter 6
Proteomic Studies of HIV-1 and Its Posttranslational Modifications

David R. Colquhoun and David R.M. Graham

Introduction: Posttranslational Modifications Important to HIV-1

HIV requires many proteins to complete its life cycle. Many of these proteins are decorated with biologically critical modifications that alter structure and function. In fact, there are many different types of posttranslational modifications (PTMs) involved in the virus life cycle (Fig. 6.1). In this chapter, we will reintroduce the virus life cycle from the perspective of PTMs. While these modifications will be described in detail later in the chapter, we will highlight new modifications in bold as they are introduced.

A mature, budded virion is a 120–150 nm diameter structure composed of a capsid (containing viral RNA, vif, vpr, nef, p7, reverse transcriptase, and integrase) and a p17 viral matrix surrounded by a lipid bilayer of host origin [1]. This lipid bilayer is studded with host proteins and the viral gp120/gp41 heterotrimer glycoprotein complex. **Glycosylation** is widely recognized to participate during the infection of target cells and in particular with interactions between gp120 and CD4. In order to infect the target cell types, $CD4^{+}$ T cells, macrophages, and microglial cells, the virus initiates binding and entry via gp120 interaction with CD4. This interaction is assisted by *N*-glycans on gp120, particularly at Asn197 [2]. This interaction with CD4 likely stabilizes the protein–protein interaction and results in a conformational change in

D.R. Colquhoun, Ph.D. (✉)
Molecular and Comparative Pathobiology, Johns Hopkins University School of Medicine, 823 Miller Research Building, 733 N. Broadway, Baltimore, MD 21205, USA
e-mail: drc@jhmi.edu

D.R.M. Graham, M.Sc., Ph.D.
Department of Molecular and Comparative Pathobiology, Johns Hopkins School of Medicine, 733 N. Broadway, MRB 835, Baltimore, MD 21205, USA
e-mail: dgraham@jhmi.edu

D.R.M. Graham, D.E. Ott (eds.), *HIV-1 Proteomics*,
DOI 10.1007/978-1-4939-6542-7_6

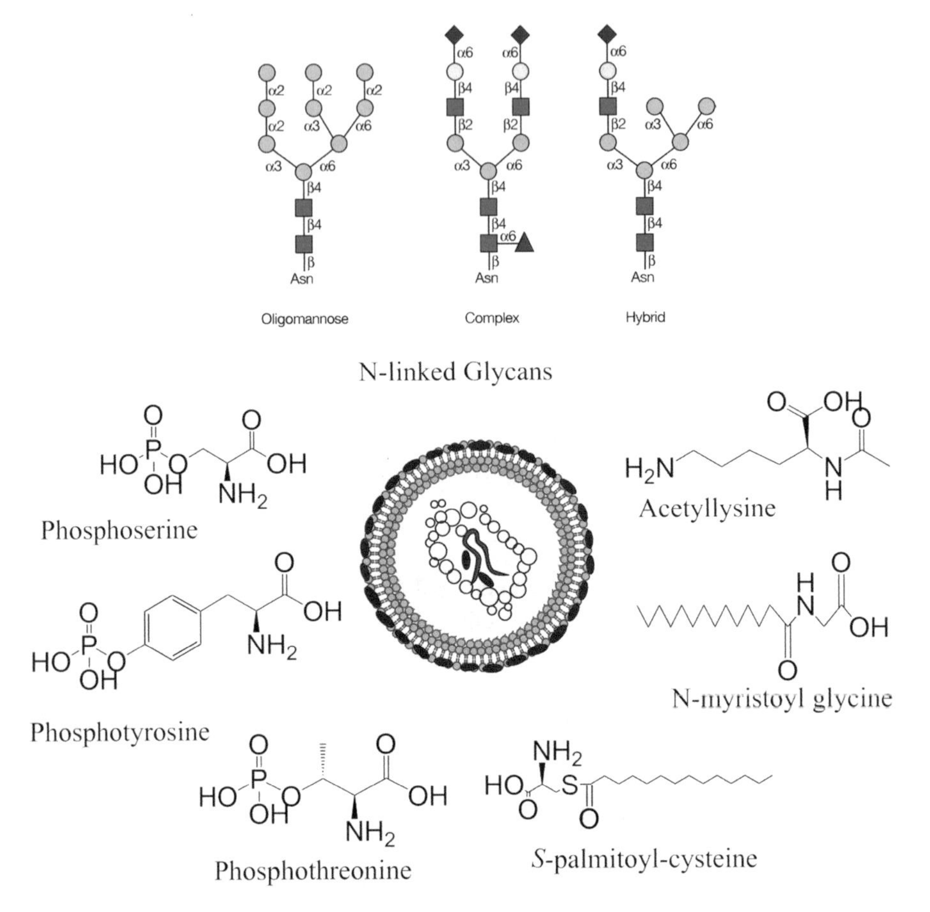

Fig. 6.1 Chemical structures of some of the major protein modifications involved in the HIV-1 life cycle. See Table 6.1 for roles and references. Glycan images modified from *Essentials of Glycobiology*, Chap. 8, Fig. 8.1

gp120 that increases the affinity for coreceptors (CCR5 and CXCR4) and exposes the **palmitoylated** protein gp41 [3]. The interactions with coreceptor further stabilize interactions exposing the fusion peptide of gp41 allowing fusion and entry of the virus [4]. Viral binding and entry can also be enhanced by dendritic cells expressing the C-type lectin (glycan-binding protein) DC-SIGN, which binds gp120 and facilitates interaction of the virus and CD4 and the coreceptor [5]. Once inside the cell, HIV deposits its payload by uncoating and initiating reverse transcription (RT).

This series of events is very poorly understood, as the events occur rapidly and components are hard to isolate and measure [6]; however, Schweitzer and colleagues have recently applied proteomics-based methodologies to understand the early state of virus uncoating and the formation of pre-integration complexes [7]. The mechanisms that drive uncoating and the early steps of infection are unknown at the level of PTMs.

The HIV cDNA generated by the RT is in complex with a number of viral proteins, including matrix and integrase, and a number of host factors related to transcription and chromatin remodeling, RNA binding, and nuclear import factors [8]. Modification of these proteins is critical to nuclear entry and integration; nuclear targeting of the process is thought to be mediated by matrix, which strongly interacts with integrase via tyrosine **phosphorylation** [9]. Conversely, it has also been reported that vpr is responsible for the karyophilic properties of the pre-integration complexes in macrophages, interacting through multiple serine phosphorylations [10]. Nevertheless, the phosphorylation of viral proteins in the pre-integration complex, thought to be carried out by protein kinase A, is critical for nuclear entry. Once in the nucleus, DNA integration must be achieved. Integrase itself is known to be **acetylated** by the host histone acetyl transferase p300, which mediates a stronger interaction between the viral genome and host genetic material and enhances the enzymatic activity of the protein. This mediates integration of the genome to "transcriptionally active regions of chromatin" [11].

Once integrated, the virus enters latency, which can last from months to years. Reactivation occurs when the viral genome is transcribed, which is initiated by activation of T cells and the increased expression of the transcription factor NF-kB. NF-kB has also been implicated in the role of acetylation in reactivation, as it is also involved in the relationship between TNFα and histone deacetylase inhibitors. As the viral mRNA is transcribed, the viral proteins rev and tat are produced, which regulate viral protein (gag and env) and RNA expression and localization [12]. The major PTMs involved in the nuclear stage of HIV infection are phosphorylation, which is active in the signal transduction and complex recruitment, and acetylation, which is a major player in transcriptional regulation. These modifications are thought to have a high frequency of cross talk and related functions [13]. A complex concert of events occurs once viral RNA and protein are produced. The gag polyprotein is co-translationally **myristoylated** in the ER on the matrix subunit, which is sequestered in a hydrophobic pocket as the polyprotein is expressed into the cytoplasm [14]. When gag either interacts with viral genomic RNA or begins to multimerize with other gag proteins, a conformational change occurs and the myristoyl group is exposed, creating a hydrophobic region that preferentially interacts with the plasma membrane [15]. Other viral factors also play a role in gag assembly [16]. Additionally, the localization is coordinated via an interaction involving the lipid tail of phosphatidylinositol 4,5-bisphosphate ($PI(4,5)P_2$) [17], which specifically targets the multimers to the lipid raft microdomains where final assembly and budding take place [18]. Meanwhile, the expressed env protein, a precursor to gp120 and gp41, trimerizes and is co-translationally **N-glycosylated** in the Golgi [19]. This gp160 is processed by furin [20] into the mature proteins, which migrate to the membrane to and assemble into a virion in lipid raft microdomains [21]. Budding then occurs at the plasma membrane and is coordinated by a number of host viral proteins, including the nucleocapsid domain of gag [22], ESCRT complexes [23], and the ubiquitin ligase nedd4 [24], which covalently attaches **ubiquitin** to p6 [25]. While the precise mechanisms are not fully understood, a picture of complex protein–protein interactions mediated by a number of critical PTMs is emerging [26].

As the virus buds, the mature virion moves to continue the life cycle by targeting lectins and receptors with the *N*-glycan decorated gp120.

The Increasing Recognition of the Role of Posttranslational Modifications in HIV-1 Proteins

The role of PTMs in HIV has been known for some time. With the advent of more powerful technology, the discoveries in this field have dramatically increased. Since 1990, the number of publications involving HIV and four of the major PTMs has increased approximately fivefold (Fig. 6.2). The areas of greatest interest are phosphorylation and glycosylation, which is logical given the frequency of these modifications (Fig. 6.3) and their important role in HIV infection and pathogenesis (Table 6.1). Other modifications have seen only modest gains in interest, likely due to the inherent challenges in studying these modifications.

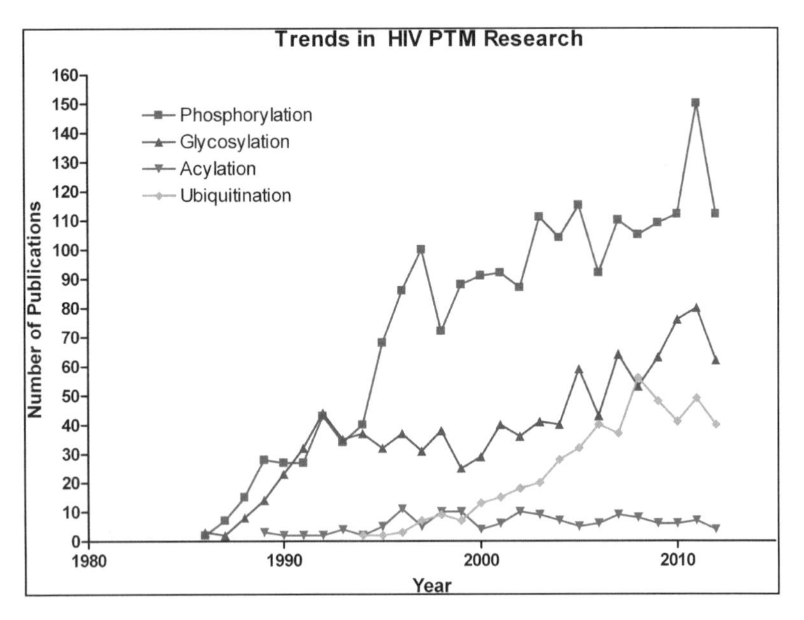

Fig. 6.2 Number of publications in HIV research describing specific posttranslational modifications. A keyword search of Scopus (http://www.scopus.com) for keywords HIV and either phosphorylation, glycosylation, acylation, or ubiquitination in the abstract, title, or keywords was carried out. Results were exported and displayed in prism

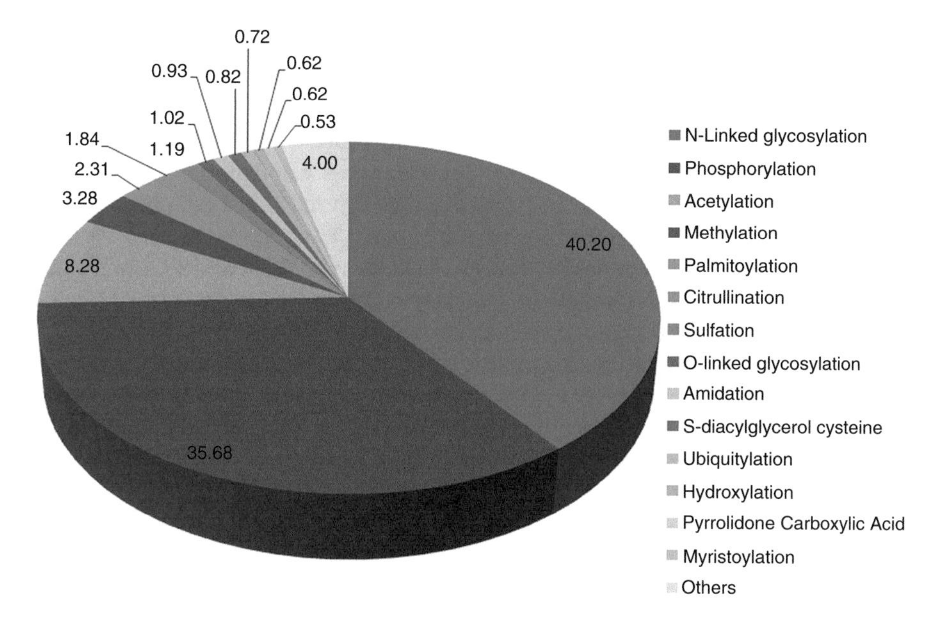

Fig. 6.3 Statistics of major posttranslational modifications on proteins. The Swiss-Prot database was mined for PTM frequency in both experimental and putitative models and the modifications enumerated [30]. Values are represented as a percentage of proteins in the entire database having at least one modification

Table 6.1 A list of some common protein modifications that affect the HIV replication cycle

Modification	Role	Site/motif	References
Phosphorylation	Signaling	S, T, Y	Francis et al. [10]
Acetylation	Transcriptional activation	K	Ott et al. [128]
N-glycosylation	Binding, immune evasion	NxS/T	Raska et al. [129]
O-glycosylation	Not known (in HIV)	S/T	Graham et al. [77]
Myristoylation	Trafficking, localization	N-terminal G	Bentham et al. [130]
Palmitoylation	Cellular localization	C	Rousso et al. [131]
Nitrosylation	Protease inactivation	C	Persichini et al. [63]
Ubiquitination	Release, degradation	K	Strack et al. [132]

Posttranslational Modifications of Host Proteins

The human genome contains somewhere in the region of 21,000 genes [27]. While alternative mRNA splicing allows for much more protein diversity by creating new protein [28], nature's answer to protein diversity appears to be posttranslational modifications. Posttranslational modifications are covalent modifications of the nascent protein following transcription and translation [29]. Posttranslational modifications are critical to protein function, localization, and protein–protein interactions. Many proteins are modified by at least one PTM; in a survey of the Swiss-Prot database, over 40 % of the sequences were predicted or experimentally observed to contain an N-glycosylation modification, and nearly as many proteins have the potential to be phosphorylated on at least one residue [30]. Other less common modifications are known to occur as well (Fig. 6.3, see uniprot.org/docs/ptmlist for a full list of PTMs).

While this serves as a good estimator of frequency, this does not account for multiple modifications on the same protein, or PTM cross talk, which is a well-established phenomenon [31, 32]. Nevertheless, it is abundantly clear that protein PTMs are common, abundant, and important regulators of function in biological systems.

Enzymatic conjugation confers some level of specificity for the target amino acid (either specific site or sequence motif), making reactions both predictable and more biologically relevant. These modification sites are determined predominantly by the specificity of the transferase or kinase utilized and may have specific amino acid motifs to specify the target amino acid sequence.

Detailed Description of PTMs Involved in the HIV-1 Life Cycle

Phosphorylation

Phosphorylation was first described by Krebs and Fischer in 1955 [33]. It involves the enzymatic addition of a phosphate (PO_4^{3-}) group to the side chain of serine, threonine, and tyrosine residues. This modification is highly dynamic and is involved in a great deal of biological processes, including tumor suppression, signal transduction, homeostasis such as insulin signaling, and protein recycling [34]. Addition of phosphate groups by kinases can activate enzymes, both by altering the tertiary structure of proteins, often resulting in changes in local hydrophobicity and exposure of active sites in the protein [35]. Regulation of phosphorylation is carried out by phosphatases and kinases, which enzymatically modify phosphorylation sites. Other forms of regulation, including cross talk with *O*-GlcNAc, have been described [31, 32]. The most thoroughly documented role of phosphorylation in HIV biology is in the nuclear host–virus interactions [10].

N-Linked Glycosylation

Attachment of complex glycans by an *N*-glycosidic bond to asparagine residues results in the production of *N*-glycans. These are, along with phosphorylation, one of the most common protein modifications observed in biology. *N*-Glycans exist in three classes: high mannose (oligomannose), complex, and hybrid [36]. They are assembled in a complex process involving many enzymes in the endoplasmic reticulum (ER) and Golgi. The functions of *N*-glycans are varied and include structural defense, self recognition, cell–cell interactions, host–pathogen interactions, regulation of enzymatic processes, and regulation of receptor-binding affinities [37]. The structures can be extremely complex, and alteration of the *N*-glycan structure can alter the function and location of the protein. It has recently been shown that protein structure can dictate the type of modification [38]. Clearly, this is a highly regulated process that is critical to cell biology; it is no surprise that HIV utilizes many of the characteristics of glycoproteins in a vast array of ways to assist in driving function, defense, and pathogenesis [39]. For example, gp120 glycosylation is proposed to act as a steric barrier or glycan shield [40] that assists the virus in evading antibody responses to gp120 protein. Evolution of the glycans confers an adaptable defense against immune response [41]. Recently, by removing these glycans, Huang and colleagues demonstrated that it is possible to generate effective neutralization using antibodies [42], reinforcing the importance of glycans in antibody evasion.

Palmitoylation and Myristoylation (Fatty Acid Acylation)

The addition of long chain lipids to proteins is termed acylation. Myristoyl group (C_{14}) chains, which are rare saturated fatty acids in cells, are covalently attached to N-terminal glycines via the enzyme *N*-myristoyl transferase (NMT) as a co-translational modification in the ER [43]. Myristoylation of proteins effects an interaction with hydrophobic regions of the cell, such as plasma membranes, allowing proteins to be localized but not necessarily anchored. Enzymatic addition of a C_{16} saturated palmitoyl group onto cysteines via a thioester linkage by palmitoyltransferases confers additional hydrophobicity. Proteins acylated in both forms will interact strongly with the inner leaflet of the plasma membrane [44]. Palmitoylation of proteins also has been shown to preferentially drive proteins to lipid raft microdomains [45]. Functionally, acylation is important for cell signaling. It is best described in its role as a modulator of G-protein-coupled receptor signaling, where dynamic modulation of palmitoylation is thought to regulate the membrane versus cytosolic location of protein domains. This regulation affects receptor activity and downstream responses in a variety of biological functions such as the endocrine, nervous, and cardiovascular systems and, more importantly in the context of HIV, immune response [46].

Ubiquitination/Ubiquitylation

Ubiquitination is the covalent addition of a small ubiquitin protein to a lysine residue in targeted proteins. The reaction is catalyzed by ubiquitin E1, E2, and E3 ligases [47]. The most well-known role of this modification is the signaling of proteins for degradation and recycling in the proteasome [48]. However, other roles have been postulated, including stress responses, DNA repair, cell cycle control, as well as viral infectivity [47]. Many of these common modifications are involved in the HIV life cycle, which hijacks and spans many common cellular pathways involved in cell signaling, cell trafficking, binding, and complex assembly. These include, but are not limited to, interactions with gag, p6, tat, and integrase, as well as signaling between viral proteins and host factors [49].

Advantages and Challenges of Studying PTMs

The study of PTMs is a burgeoning field. As our overall understanding, biological processes become more detailed; the roles of PTMs in many diseases, cellular functions, and regulation have been revealed [50]. The role of dynamic modifications, such as phosphorylation and glycosylation, was initially characterized by mutation of specific sites of modification, studied using well-characterized protein models. However, the use of -omics technologies has allowed the high-throughput analysis of entire cell cultures or tissues for specific modifications [51]. In addition, more recent work indicates some of these modifications may act in concert, and the dynamic interplay between modifications can dramatically affect the biological state of a system [31, 32]. Finally, since the occupancy or proportion of modified proteins can play a role in the biological effect, novel analytical techniques are being developed in order to characterize and both absolutely and relatively quantify the amount of modified proteins present in different biological states. With the advancement of PTM studies and techniques, the need for comprehensive, reproducible, and validated techniques has become clearly apparent. As with any scientific experiment, advantages, challenges, and limitations must be understood in order to comprehend and place any findings into a biological context.

The single most important advantage of studying PTMs is that we are able to understand a more detailed biological picture of the system being studied. In the context of HIV, by merely measuring protein levels in a cell, the mechanism for the subcellular localization of viral proteins would not be well understood. In contrast, the study of PTMs allows for better characterization, measurement, or emulation of a particular biological system, which is the goal of basic science experiments. The recent interest in systems, or multi-omic, studies emphasizes the need to place findings in the context of all the components [52]. This in itself presents challenges and limitations, since in many instances the comprehensive measurement of proteins, lipids, nucleic acids, and metabolites is untenable or prohibitively expensive to undertake. However, studies that successfully integrate different disciplines often reveal novel mechanisms and associations that were otherwise unapparent. For example, the role of the inner membrane PI(4,5)P_2 in gag multimerization and membrane localization required a multidisciplinary approach to clearly identify biological function and role in the process [53, 54].

Many PTMs, as previously noted, are dynamic processes that change in response to stimulus, such as immune activation, stress responses, or analyte and substrate levels. Developing an understanding of the role of dynamic modulation of proteins can lead to a striking view of cellular regulation. For example, the role of *O*-GlcNAc in a large number of diseases is only starting to be realized, and the cross talk between glycosylation and phosphorylation is revealing novel disease markers and mechanisms [55]. Many of these processes initialize a chain reaction of downstream effects, massively altering gene and protein expression in a cell, such as in the use of *N*-myristoyl transferase by Nef in HIV infection [56]. As diseases progress, the levels of modifications may change dramatically. By quantifying PTM levels, the progress or extent of a disease may be better characterized and measured [57]. Subtle interventions at specific targets may be sufficient to abrogate infectivity of viruses [58], to prevent completion of the viral life cycle [59], or to better elucidate poorly understood phases such as latency [60].

Although there are a number of benefits to studying PTMs, there are also clear challenges. Some of these are presented by limitations in instrumentation, such as preservation of modifications and their subsequent detection, which will be covered in detail later in the chapter. Other limitations are inherent in the biology of PTMs, such as temporal expression and modulation, dynamic range, and more broadly, the a priori biological understanding of how subtly modifying a side chain residue of a single amino acid can affect signal downstream biological effects.

Proteomic Methods for PTM Analysis

In order to measure PTMs from biological samples, specific a priori decisions must be made in order to select the correct technique, sample preparation, and analysis strategies. These decisions typically require an assessment of which modifications are being studied and the specificity and range of the involved proteins. For example, if S-palmitoylation is a target, biochemical membrane isolation should be carried out as the proteins of interest are most likely present in this cellular fraction [45]. Conversely, when studying protein phosphorylation, specific inhibitors should be supplemented in the extraction buffer, and sample perturbation should be minimized [61]. Following sample extraction, the choice of analytical technique is often driven by availability of equipment and samples and the cost of an assay. A number of biochemical, gel-based and MS techniques are available, and these are outlined below.

Biochemical Methods

One of the major challenges of studying some PTMs is the lability of the modification and thus the biological interpretation of any analytical data. As discussed earlier, thoughtful and careful sample preparation with knowledge of the sample lability and treatment conditions can vastly improve the experimental outcomes. In some circumstances, the relative reactivity of modifications can be used in chemical labeling strategies. One example is the detection and measurement of S-nitrosylation, a modification thought to be important in cell signaling, immune defense, and pathogenesis of some diseases [62]. This modification has been shown to have antiviral activity, in part due to its ability to inactivate the HIV-1 protease [63]. The modification itself is extremely susceptible to oxidation and is light sensitive, making measurement difficult. Using the differential oxidation to their advantage, Jaffrey et al. developed a biotin switch method whereby mild ascorbate treatment released the nitrosylation, and the thiol was biotin labeled and subsequently captured [64]. Further refinements have led to fluorescent labeling techniques [65] and other MS-compatible applications [66, 67]. Other dynamic modifications, such as *O*-GlcNAc, can be similarly modified and subsequently measured [68].

Difference Gel Electrophoresis and 2D Electrophoresis

Two-dimensional gel electrophoresis (2DE) was first described by O'Farrell and Klose [69, 70]. With the advent of fluorescent protein labeling, differential analysis within individual gels became possible [71]. Difference gel electrophoresis (DIGE) and 2DE have been used extensively for many years in HIV research (e.g., [72–75]). However, with the development of robust LC-MS/MS techniques, the application of DIGE and gel-based protocols has declined. This is primarily due to the complexity of data analysis and spot identification and the great deal of time and investment required, compared to an iTRAQ or other quantitative MS approach. Furthermore, protein identifications are made following quantification, so only those proteins deemed of interest are actually identified by mass spectrometry (MS). The amount of protein required to visualize low-abundance proteins can be quite high, and reproducibility can be an issue.

With all that in mind, there are several "niche" applications that 2DE and in particular DIGE are well suited to. One of these is the analysis of PTMs. Using two-dimensional electrophoresis followed by Western blotting, Davis et al. were able to demonstrate that HIV-1 reverse transcriptase is present in a number of protein isoforms within the cell [76]. The isoforms were shown by phosphatase activity to be in part due to phosphorylation, and additional experiments showed that the majority of the protein was in the modified form. The phosphorylation site was not determined, as MS analysis was not carried out. In another study, Graham and coworkers described the use of DIGE to differentially identify PTMs in HIV and SIV [77]. By enzymatically cleaving *N*-glycans from the virus in one of two samples, and differentially labeling the resulting proteins, modified proteins were easily visualized, as spot locations changed for deglycosylated glycoproteins. Taking advantage of the deamidation of N to D in PNGase F treatment, MS was utilized to identify sites of N-glycosylation on gp120 for both HIV and SIV. This approach is advantageous, as modified proteins are easily visualized and can be excised for MS identification.

Although not carried out in this study, 2DE samples are also compatible with glycan isoform analysis (see below). Using this technique, along with well-defined hypotheses, allows proteomic screening for specific PTMs of interest and gives information on relative proportions of modified versus unmodified proteins.

Mass Spectrometry of PTMs

Mass spectrometry has been applied to proteins for several decades [78, 79]. "Bottom-up" MS, where tryptic peptides are analyzed and searched against genomic databases [80], has been used with much success for the last decade [31, 32, 81, 82]. Recently, technological advances have rapidly improved the sensitivity, selectivity, and resolution of instrumentation [83, 84]. Naturally, this improvement in technology has enabled an emphasis on targeting specific peptides, and subsequently PTMs, since one of the overarching goals of many proteomic studies is to apply technology to biological questions and place these studies in a biologically relevant framework [52], in which PTMs play a critical role. Thus, in addition to the technical gains in instrumentation, novel methods have been applied to the fragmentation and detection of biomolecules in MS. These methods include MS^3 and MS^n analysis, as well as alternative fragmentation methods such as electron capture dissociation/electron transfer dissociation (ECD/ETD) and higher-energy collision dissociation (HCD) in the orbitrap [85].

Fragmentation of some modified peptides by collision-induced dissociation (CID), particularly phosphorylation and GlcNAcylation, results in elimination of the modification and poor fragmentation on the peptide backbone, typically reporting a mass of the precursor minus the modification, or a neutral loss of m/z 80 and 97 (phosphorylation of Y and S/T, respectively) or 203 (hexosamine). In trapping instruments such as ion traps (which, as the name implies, can accumulate ions until a specified threshold is attained), this neutral loss event was compensated for in one of two ways: firstly, carry out data-dependent neutral loss (DDNL) analysis, whereby a neutral loss of specific m/z in MS to MS^2 triggered an accumulation of the neutral loss ion and subsequent MS^3 fragmentation [86], or, secondly, trigger a second MS^n event while still accumulating data, which is termed "pseudoMS^n" [87]. This development was particularly important since up to an estimated 80 % of ions from a phosphopeptide enrichment contained neutral loss ions [86].

In addition to MS^3 approaches, alternative fragmentation methods have been developed. To overcome the energy transfer to the PTM and poor fragmentation, ECD was developed for Fourier transform MS [88]. This utilizes a free electron, which reacts with the peptide backbone in an exothermic manner, resulting in fragmentation of the backbone without affecting proximal phospho groups [89]. In ETD, the same principle applies, but the reaction is catalyzed by an anthracene anion in ion trap or orbitrap instruments [90]. The resulting MS/MS spectra are rich in *c* and *z* ions and give a much better sequence coverage than CID for modified peptides. One important caveat, however, is that ETD has been shown to require larger, higher charge state precursors for optimal results. Recent work has developed a more robust workflow using the endoprotease LysC in combination with some mobile phase modifiers to optimize results [91]. To our knowledge, these techniques have not been applied to HIV PTM study, making this a potentially fruitful area of research.

Advanced MS Techniques

The data-dependent acquisition utilized by bottom-up MS creates challenges. While bottom-up MS is suited to proteomic discovery, it is not optimal for PTM analysis. As previously mentioned, the detection of PTMs is reliant on the a priori knowledge of their presence. In addition, searching for multiple PTMs can dramatically increase the time taken for data analysis. One way to overcome those limitations is to utilize data-*independent* acquisition. This technique goes by a number of names: SWATH-MS, MS^E, AIF, or PAcIFIC [92]. In this method, the instrument scans mass ranges in MS/MS without isolating specific precursor ions, essentially capturing all the ions present within a given dynamic range. The advantage of this is that you can screen for PTMs based upon a given mass change; the disadvantage is that these experiments are data independent so any neutral loss ions cannot be further fragmented for structural or compositional information.

Another method that has garnered recent attention is the use of "top-down" MS. In contrast to bottom-up, the intact protein is introduced to the mass spectrometer, and subsequently fragmented into smaller pieces, which are then analyzed for the presence of modifications [93]. While it is extremely powerful and can identify any known modification (complex *N*-glycans and other complex modifications aside), there are limitations. The protein sequence must be known in order to fit the fragments to the sequence, and the protein must be purified and of a size that is currently amenable to analysis (large proteins do not sufficiently ionize). While a relatively robust sample preparation tool exists [94], the challenges of implementing this technology still need to be overcome. Once established, this may change the way PTMs are studied in many research areas. For HIV, the ability to screen viruses from different cell types may help to establish in more detail mechanisms of viral tropism and immune responses to different viruses.

Emerging Technologies

A number of emerging technologies are enabling a deeper and more directed analysis of HIV and host PTMs. Advanced sample preparation and labeling strategies will allow for refined analysis and capture of specifically modified proteins, and instrumental and analytical updates will enable analysis of PTMs and proteins from novel and beneficial angles. The two main areas of these emerging technologies are sample labeling/enrichment and improved analytical technologies such as instrumentation and bioinformatic strategies. From the labeling and enrichment side, the emergence of click chemistry tools and techniques has the potential to dramatically change the way that we study PTMs, since it removes the barriers and limitations of the aforementioned radiolabeling, and is more specific than chromatographic enrichment and analysis. From an analytical aspect, the explosion of simple, accessible nonprotein MS methods, and the subsequent technological advances, opens the door to analyzing not only the proteome but also the functional modifications attached to the proteins in a site-specific manner. Finally, the utilization of the nascent technology of MS imaging, while currently limited in application and robustness, may provide a tool for the specific localization of proteins and their modified counterparts in tissues and cells.

Click Chemistry

Click chemistry is the use of bio-orthogonal synthetic compounds for the specific labeling and capture of modified residues. These residues can be specifically labeled in order to study a particular PTM on proteins or a specific cellular pathway. The term "click chemistry" was coined by Barry Sharpless, who described it as "spring-loaded" reactions, "destined for a single trajectory" [95]. These reactions typically involve 1,3-dipolar cycloadditions that react an azide to an alkyne functional group either catalyzed by copper [96] or using copper-free chemistry for in vivo studies [97]. There are a number of commercially available azide- and alkyne-containing substrates. A great deal of the seminal work was carried out studying cell surface *O*-glycans by feeding azide-modified precursors to cells and detecting incorporation via Western blotting to a clicked substrate, e.g., FLAG tag [96]. Other substrates for modification include *N*-glycans such as sialic acid and fucose precursors [98, 99]; acylations such as palmitoylation, myristoylation, and prenylation [100, 101]; and newly synthesized proteins using azidohomoalanine [102, 103]. As we understand more about the mechanisms and chemistry of PTM incorporation, improvements are made in the selectivity and sensitivity of chemical analogs. For example, by modifying the permeability of compounds, it is possible to reduce the dose of analog to achieve equal labeling efficiency, reducing potential off-target effects [98]. Furthermore, by altering the initial "click" substrate, one may change the target specificity and further define specific subpopulations of proteins and their modifications.

In HIV research, as discussed above, the role of specific PTMs in the life cycle, both on the host and viral sides, still remains largely unknown. However, the effects of some more studied modifications are beginning to be elucidated. For example, it is well known that surface glycosylation of viral proteins is critical in evading and adapting to the host immune response [42, 104, 105]. Obviously, as a high number of cell surface proteins are glycosylated, glycan–protein interactions are known to be critical for cell–cell communication, as well as HIV–cell interactions [2, 106]. These primary interactions are critical and provide insight into specific viral–host mechanistic relationships; however, by specifically targeting a subpopulation of proteins, we may yield more information regarding more subtle secondary interactions that may be just as critical to the viral life cycle [107].

Non-proteomic Mass Spectrometry

The emergence of proteomic MS in the last 30 years has led to an explosion in discovery at the protein level. The methods of bottom-up MS and data analysis using proteogenomic databases have allowed broad access to understanding the "what," "when," and relatively "how much" questions asked of protein expression in specific systems and disease states. In HIV, this has resulted in a broader understanding of the proteins that play a role in HIV infection and life cycles [31, 32, 81, 82, 108–110], as well as the minimally conserved proteome of HIV itself [107, 111]. These studies have certainly advanced knowledge in the field of HIV biology; however, a challenge associated with this type of research is that more subtle biology may be missed, considering we are only studying proteotypic peptides with defined modifications, that is, by searching against protein and genomic databases, and by selecting certain analysis tools and methods, we choose to ignore a vast wealth of information about protein modifications and diversity. For example, labile modifications such as phosphorylations may be lost during sample preparation or analysis, and while a protein may be defined as critical, it may be the effect of phosphorylation that defines the biology. While methods for phosphoproteomic analysis and other relatively simple modifications (e.g., *O*-GlcNAc) are developing rapidly [51, 112, 113], other modifications are not so straightforwardly analyzed. The use of ECD/ETD fragmentation has allowed site-specific analysis of glycoproteins [114], but the detection of glycan isoforms remains a very specialized technique. As mentioned above, a single protein may be multiply glycosylated throughout the sequence, and each site of glycosylation can contain a large number of glycan isoforms, some of which may be functionally distinct [115, 116]. Methods for glycan isoform identification and site localization may lead to a greater understanding of the roles of oligosaccharides on HIV attachment, virion–cell interaction, infectivity, and the host immune response.

The methods for releasing *N*-glycans are well established, using enzymatic cleavage of the glycan via peptide-N4-(*N*-acetyl-beta-glucosaminyl)asparagine amidase (PNGase F) [117]. However, separation and analysis of the glycan isoforms are complicated by relative homogeneity of the saccharide monomers. Non-LC methods for analysis are primarily driven by MALDI applications, where spotted glycans are identified predominantly by their precursor mass, and some structural information is yielded by MS/MS analysis [118]. This method does limit analysis to higher-abundance glycans, and LC separation methods have been advanced in the last several years to compensate for this. One recent advance is the use of porous graphitized carbon (PGC) HPLC-Chip columns for nano-LC separation of glycans and their subsequent analysis by ESI-MS/MS. In this method, the *N*-glycans are released from the protein backbone, enriched using solid-phase extraction, and separated on a gradient of water/acetonitrile. Glycan structure is assigned based upon the precursor mass and using MS/MS fragmentation information; the numbers of hexose, hexNAc, fucose, NeuAc, and NeuGc monomers and some unambiguous structures are assigned [119]. While powerful, this technique does not allow the connection between the protein structure and the glycan. In order to achieve this, one must isolate both a proteotypic peptide sequence *and* the modification in the same MS/MS spectrum. This has been successfully applied by using pronase to nonspecifically digest the protein backbone, leaving short amino acid tags on the glycans, which are then subjected to LC-MS/MS. This technique relies on the a priori knowledge of the protein sequence. However, if this is known from a complementary analysis, the site occupancy of glycan isoforms on a specific protein may be elucidated [120, 121]. This structural information could have a myriad of applications to HIV biology, from both the host and virion aspect. In conjunction with non-MS methods (e.g., lectin arrays), more detailed understanding of the gp120 glycan shield may allow better design or specificity for arrays of neutralizing antibodies and knowledge of the specific cell–virus interactions [2, 122]. Glycans may also play a role in viral tropism, and identifying host–virus interactions to specific glycoproteins may lead to novel approaches to clinical diagnosis and interventions.

Mass Spectrometry Imaging

While non-proteomic MS techniques are rapidly advancing, there are additional MS tools that are beginning to emerge as potentially powerful tools for the analysis of biological systems. Mass spectrometry imaging (MSI) is a technique that has shown much promise in the past several years. Although MSI was primarily developed for the analysis of small molecules and lipids in tissues, much time has been invested into developing robust methods for the analysis of proteins and peptides in order to add a spatial dimensionality to proteomics information. Analysis of small proteins [123] and peptides [124, 125] is becoming more routine, and novel methods such as 3D imaging are being proposed [126]. However, the analysis of PTMs is some way off, and to our knowledge there are no publications in this area, suggesting an area for further exploration and expansion. Although some limited work on glycan monomers and simple carbohydrates has demonstrated the potential efficacy of this technique for glycan analysis [127], robust and reproducible methods must be established before any biological studies can be launched. The potential to localize specific modified proteins and peptides to tissue and biological regions could be applied to understanding the role of HIV infection in systemic disorders such as HIV-associated neurocognitive dysfunction.

Summary

The analysis of the role of PTMs in HIV is growing, and our understanding of the basic biological roles of modifications on the life cycle of the virus is increasing. With the development of enhanced analytical techniques and better mechanisms for specific labeling, enrichment, and analysis of modifications, it is anticipated that further developments will be made in understanding the critical roles of PTMs. In particular, the interplay of lipid modifications for trafficking proteins to the cell membrane coupled with glycosylations critical to cellular communication, binding, and interaction may yield a greater understanding of the roles of host proteins integrated into HIV and reveal potential mechanisms of disrupting the viral life cycle. In addition, a thorough characterization of the modifications of both host and viral proteins present in HIV will allow researchers to develop a better systems biology picture of the virus, and potentially connect viral composition to previously unrelated cellular pathways, revealing novel mechanisms for intervention and drug targeting. Further, by continuing the characterization of the glycome of the virus, and in particular gp120, the enhancement of potentially neutralizing vaccines could be realized. In conclusion, the burgeoning field of posttranslational protein modifications will continue to benefit our understanding of HIV and its clinical effects.

References

1. Briggs JA, et al. Structural organization of authentic, mature HIV-1 virions and cores. EMBO J. 2003;22(7):1707–15.
2. Wilhelm D, et al. Glycosylation assists binding of HIV protein gp120 to human CD4 receptor. Chembiochem. 2012;13(4):524–7.
3. Bhattacharya J, et al. Human immunodeficiency virus type 1 envelope glycoproteins that lack cytoplasmic domain cysteines: impact on association with membrane lipid rafts and incorporation onto budding virus particles. J Virol. 2004;78(10):5500–6.
4. Wilen CB, et al. Molecular mechanisms of HIV entry. Adv Exp Med Biol. 2012;726:223–42.
5. Geijtenbeek TB, et al. DC-SIGN, a dendritic cell-specific HIV-1-binding protein that enhances trans-infection of T cells. Cell. 2000;100(5):587–97.
6. Fassati A. Multiple roles of the capsid protein in the early steps of HIV-1 infection. Virus Res. 2012;170(1–2):15–24.
7. Schweitzer CJ, et al. Proteomic analysis of early HIV-1 nucleoprotein complexes. J Proteome Res. 2013;12(2):559–72.
8. Allouch A, Cereseto A. Identification of cellular factors binding to acetylated HIV-1 integrase. Amino Acids. 2011;41(5):1137–45.
9. Gallay P, et al. HIV nuclear import is governed by the phosphotyrosine-mediated binding of matrix to the core domain of integrase. Cell. 1995;83(4):569–76.
10. Francis AC, et al. Role of phosphorylation in the nuclear biology of HIV-1. Curr Med Chem. 2011;18(19):2904–12.
11. Cereseto A, et al. Acetylation of HIV-1 integrase by p300 regulates viral integration. EMBO J. 2005;24(17):3070–81.
12. Van Lint C, et al. Molecular mechanisms involved in HIV-1 transcriptional latency and reactivation: implications for the development of therapeutic strategies. Bull Mem Acad R Med Belg. 2004;159(Pt 2):176–89.
13. Yang XJ, Seto E. Lysine acetylation: codified crosstalk with other posttranslational modifications. Mol Cell. 2008;31(4):449–61.
14. Provitera P, et al. The effect of HIV-1 Gag myristoylation on membrane binding. Biophys Chem. 2006;119(1):23–32.
15. Tang C, et al. Entropic switch regulates myristate exposure in the HIV-1 matrix protein. Proc Natl Acad Sci U S A. 2004;101(2):517–22.

16. Morikawa Y, et al. Roles of matrix, p2, and N-terminal myristoylation in human immunodeficiency virus type 1 Gag assembly. J Virol. 2000;74(1):16–23.
17. Saad JS, et al. Structural basis for targeting HIV-1 Gag proteins to the plasma membrane for virus assembly. Proc Natl Acad Sci U S A. 2006;103(30):11364–9.
18. Nguyen DH, Hildreth JE. Evidence for budding of human immunodeficiency virus type 1 selectively from glycolipid-enriched membrane lipid rafts. J Virol. 2000;74(7):3264–72.
19. Murakami T. Roles of the interactions between Env and Gag proteins in the HIV-1 replication cycle. Microbiol Immunol. 2008;52(5):287–95.
20. Moulard M, et al. Processing and routage of HIV glycoproteins by furin to the cell surface. Virus Res. 1999;60(1):55–65.
21. Checkley MA, et al. HIV-1 envelope glycoprotein biosynthesis, trafficking, and incorporation. J Mol Biol. 2011;410(4):582–608.
22. Dussupt V, et al. Basic residues in the nucleocapsid domain of Gag are critical for late events of HIV-1 budding. J Virol. 2011;85(5):2304–15.
23. Usami Y, et al. The ESCRT pathway and HIV-1 budding. Biochem Soc Trans. 2009;37(Pt 1):181–4.
24. Sette P, et al. The ESCRT-associated protein Alix recruits the ubiquitin ligase Nedd4-1 to facilitate HIV-1 release through the LYPXnL L domain motif. J Virol. 2010;84(16):8181–92.
25. Gurer C, et al. Covalent modification of human immunodeficiency virus type 1 p6 by SUMO-1. J Virol. 2005;79(2):910–7.
26. Sundquist WI, Krausslich HG. HIV-1 assembly, budding, and maturation. Cold Spring Harb Perspect Med. 2012;2(7):a006924.
27. Pennisi E. Genomics. ENCODE project writes eulogy for junk DNA. Science. 2012;337(6099):1159, 1161.
28. Black DL. Mechanisms of alternative pre-messenger RNA splicing. Annu Rev Biochem. 2003;72:291–336.
29. Wold F. In vivo chemical modification of proteins (post-translational modification). Annu Rev Biochem. 1981;50:783–814.
30. Khoury GA, et al. Proteome-wide post-translational modification statistics: frequency analysis and curation of the swiss-prot database. Sci Rep. 2011;1:90.
31. Wang T, et al. HIV-1-infected astrocytes and the microglial proteome. J Neuroimmune Pharmacol. 2008;3(3):173–86.
32. Wang Z, et al. Cross-talk between GlcNAcylation and phosphorylation: site-specific phosphorylation dynamics in response to globally elevated O-GlcNAc. Proc Natl Acad Sci U S A. 2008;105(37):13793–8.
33. Krebs EG, Fischer EH. Phosphorylase activity of skeletal muscle extracts. J Biol Chem. 1955;216(1):113–20.
34. Manning G, et al. Evolution of protein kinase signaling from yeast to man. Trends Biochem Sci. 2002;27(10):514–20.
35. Barford D, et al. The structure and mechanism of protein phosphatases: insights into catalysis and regulation. Annu Rev Biophys Biomol Struct. 1998;27:133–64.
36. Stanley P, et al. N-Glycans. Cold Spring Harbor: Cold Spring Harbor Laboratory Press; 2009.
37. Varki A, Lowe JB. Biological roles of glycans. Cold Spring Harbor: Cold Spring Harbor Laboratory Press; 2009.
38. Thaysen-Andersen M, Packer NH. Site-specific glycoproteomics confirms that protein structure dictates formation of N-glycan type, core fucosylation and branching. Glycobiology. 2012;22(11):1440–52.
39. Sato S, et al. Glycans, galectins, and HIV-1 infection. Ann N Y Acad Sci. 2012;1253(1):133–48.
40. Wei X, et al. Antibody neutralization and escape by HIV-1. Nature. 2003;422(6929):307–12.
41. Dacheux L, et al. Evolutionary dynamics of the glycan shield of the human immunodeficiency virus envelope during natural infection and implications for exposure of the 2G12 epitope. J Virol. 2004;78(22):12625–37.

42. Huang X, et al. Highly conserved HIV-1 gp120 glycans proximal to CD4-binding region affect viral infectivity and neutralizing antibody induction. Virology. 2012;423(1):97–106.
43. Boutin JA. Myristoylation. Cell Signal. 1997;9(1):15–35.
44. van't Hof W, Resh MD. Targeting proteins to plasma membrane and membrane microdomains by N-terminal myristoylation and palmitoylation. Methods Enzymol. 2000;327:317–30.
45. Levental I, et al. Greasing their way: lipid modifications determine protein association with membrane rafts. Biochemistry. 2010;49(30):6305–16.
46. Wettschureck N, Offermanns S. Mammalian G proteins and their cell type specific functions. Physiol Rev. 2005;85(4):1159–204.
47. Finley D, Chau V. Ubiquitination. Annu Rev Cell Biol. 1991;7:25–69.
48. Wilkinson KD. Ubiquitination and deubiquitination: targeting of proteins for degradation by the proteasome. Semin Cell Dev Biol. 2000;11(3):141–8.
49. Biard-Piechaczyk M, et al. HIV-1, ubiquitin and ubiquitin-like proteins: the dialectic interactions of a virus with a sophisticated network of post-translational modifications. Biol Cell. 2012;104(3):165–87.
50. Li S, et al. Loss of post-translational modification sites in disease. Pac Symp Biocomput. 2010;8:337–47.
51. Macek B, et al. Global and site-specific quantitative phosphoproteomics: principles and applications. Annu Rev Pharmacol Toxicol. 2009;49:199–221.
52. Hood L. Systems biology: integrating technology, biology, and computation. Mech Ageing Dev. 2003;124(1):9–16.
53. Campbell S, et al. Modulation of HIV-like particle assembly in vitro by inositol phosphates. Proc Natl Acad Sci U S A. 2001;98(19):10875–9.
54. Ono A, et al. Phosphatidylinositol (4,5) bisphosphate regulates HIV-1 Gag targeting to the plasma membrane. Proc Natl Acad Sci U S A. 2004;101(41):14889–94.
55. Mishra S, et al. O-GlcNAc modification: why so intimately associated with phosphorylation? Cell Commun Signal. 2011;9(1):1.
56. Morita D, et al. Cutting edge: T cells monitor N-myristoylation of the Nef protein in simian immunodeficiency virus-infected monkeys. J Immunol. 2011;187(2):608–12.
57. Zhang J, et al. Top-down quantitative proteomics identified phosphorylation of cardiac troponin I as a candidate biomarker for chronic heart failure. J Proteome Res. 2011;10(9):4054–65.
58. Platt EJ, et al. Kinetic mechanism for HIV-1 neutralization by antibody 2G12 entails reversible glycan binding that slows cell entry. Proc Natl Acad Sci U S A. 2012;109(20):7829–34.
59. Saermark T, et al. Characterization of N-myristoyl transferase inhibitors and their effect on HIV release. AIDS. 1991;5(8):951–8.
60. Budhiraja S, et al. Cyclin T1 and CDK9 T-loop phosphorylation are downregulated during establishment of HIV-1 latency in primary resting memory CD4+ T cells. J Virol. 2013;87(2):1211–20.
61. Kanshin E, et al. Sample preparation and analytical strategies for large-scale phosphoproteomics experiments. Semin Cell Dev Biol. 2012;23(8):843–53.
62. Colasanti M, et al. S-nitrosylation of viral proteins: molecular bases for antiviral effect of nitric oxide. IUBMB Life. 1999;48(1):25–31.
63. Persichini T, et al. Cysteine nitrosylation inactivates the HIV-1 protease. Biochem Biophys Res Commun. 1998;250(3):575–6.
64. Jaffrey SR, et al. Protein S-nitrosylation: a physiological signal for neuronal nitric oxide. Nat Cell Biol. 2001;3(2):193–7.
65. Santhanam L, et al. Selective fluorescent labeling of S-nitrosothiols (S-FLOS): a novel method for studying S-nitrosation. Nitric Oxide. 2008;19(3):295–302.
66. Kohr MJ, et al. Measurement of s-nitrosylation occupancy in the myocardium with cysteine-reactive tandem mass tags: short communication. Circ Res. 2012;111(10):1308–12.

67. Murray CI, Van Eyk JE. A twist on quantification: measuring the site occupancy of s-nitrosylation. Circ Res. 2012;111(10):1253–5.
68. Wells L, et al. Mapping sites of O-GlcNAc modification using affinity tags for serine and threonine post-translational modifications. Mol Cell Proteomics. 2002;1(10):791–804.
69. Klose J. Protein mapping by combined isoelectric focusing and electrophoresis of mouse tissues. A novel approach to testing for induced point mutations in mammals. Humangenetik. 1975;26(3):231–43.
70. O'Farrell PH. High resolution two-dimensional electrophoresis of proteins. J Biol Chem. 1975;250(10):4007–21.
71. Unlu M, et al. Difference gel electrophoresis: a single gel method for detecting changes in protein extracts. Electrophoresis. 1997;18(11):2071–7.
72. Kramer G, et al. Proteomic analysis of HIV-T cell interaction: an update. Front Microbiol. 2012;3:240.
73. Melendez LM, et al. Proteomic analysis of HIV-infected macrophages. J Neuroimmune Pharmacol. 2011;6(1):89–106.
74. Pocernich CB, et al. Proteomics analysis of human astrocytes expressing the HIV protein Tat. Brain Res Mol Brain Res. 2005;133(2):307–16.
75. Rozek W, et al. Cerebrospinal fluid proteomic profiling of HIV-1-infected patients with cognitive impairment. J Proteome Res. 2007;6(11):4189–99.
76. Davis AJ, et al. Human immunodeficiency virus type-1 reverse transcriptase exists as post-translationally modified forms in virions and cells. Retrovirology. 2008;5:115.
77. Graham DR, et al. Two-dimensional gel-based approaches for the assessment of N-Linked and O-GlcNAc glycosylation in human and simian immunodeficiency viruses. Proteomics. 2008;8(23–24):4919–30.
78. Fenn JB, et al. Electrospray ionization for mass spectrometry of large biomolecules. Science. 1989;246(4926):64–71.
79. Tanaka K, et al. Protein and polymer analyses up to m/z 100 000 by laser ionization time-of-flight mass spectrometry. Rapid Commun Mass Spectrom. 1988;2(8):151–3.
80. McCormack AL, et al. Direct analysis and identification of proteins in mixtures by LC/MS/MS and database searching at the low-femtomole level. Anal Chem. 1997;69(4):767–76.
81. Coiras M, et al. Modifications in the human T cell proteome induced by intracellular HIV-1 Tat protein expression. Proteomics. 2006;6 Suppl 1:S63–73.
82. Zhang L, et al. Host proteome research in HIV infection. Genomics Proteomics Bioinformatics. 2010;8(1):1–9.
83. Dillen L, et al. Comparison of triple quadrupole and high-resolution TOF-MS for quantification of peptides. Bioanalysis. 2012;4(5):565–79.
84. Olsen JV, et al. A dual pressure linear ion trap Orbitrap instrument with very high sequencing speed. Mol Cell Proteomics. 2009;8(12):2759–69.
85. Singh C, et al. Higher energy collision dissociation (HCD) product ion-triggered electron transfer dissociation (ETD) mass spectrometry for the analysis of N-linked glycoproteins. J Proteome Res. 2012;11(9):4517–25.
86. Villen J, et al. Evaluation of the utility of neutral-loss-dependent MS3 strategies in large-scale phosphorylation analysis. Proteomics. 2008;8(21):4444–52.
87. Schroeder MJ, et al. A neutral loss activation method for improved phosphopeptide sequence analysis by quadrupole ion trap mass spectrometry. Anal Chem. 2004;76(13):3590–8.
88. Stensballe A, et al. Electron capture dissociation of singly and multiply phosphorylated peptides. Rapid Commun Mass Spectrom. 2000;14(19):1793–800.
89. Syka JE, et al. Peptide and protein sequence analysis by electron transfer dissociation mass spectrometry. Proc Natl Acad Sci U S A. 2004;101(26):9528–33.
90. Coon JJ, et al. Electron transfer dissociation of peptide anions. J Am Soc Mass Spectrom. 2005;16(6):880–2.
91. Xie LQ, et al. Improved proteomic analysis pipeline for LC-ETD-MS/MS using charge enhancing methods. Mol Biosyst. 2012;8(10):2692–8.

92. Gillet LC, et al. Targeted data extraction of the MS/MS spectra generated by data-independent acquisition: a new concept for consistent and accurate proteome analysis. Mol Cell Proteomics. 2012;11(6):O111.016717.
93. Tran JC, et al. Mapping intact protein isoforms in discovery mode using top-down proteomics. Nature. 2011;480(7376):254–8.
94. Lee JE, et al. A robust two-dimensional separation for top-down tandem mass spectrometry of the low-mass proteome. J Am Soc Mass Spectrom. 2009;20(12):2183–91.
95. Kolb HC, et al. Click chemistry: diverse chemical function from a few good reactions. Angew Chem Int Ed Engl. 2001;40(11):2004–21.
96. Kiick KL, et al. Incorporation of azides into recombinant proteins for chemoselective modification by the Staudinger ligation. Proc Natl Acad Sci U S A. 2002;99(1):19–24.
97. Chang PV, et al. Copper-free click chemistry in living animals. Proc Natl Acad Sci U S A. 2010;107(5):1821–6.
98. Almaraz RT, et al. Metabolic oligosaccharide engineering with N-Acyl functionalized ManNAc analogs: cytotoxicity, metabolic flux, and glycan-display considerations. Biotechnol Bioeng. 2012;109(4):992–1006.
99. Hart C, et al. Metabolic labeling and click chemistry detection of glycoprotein markers of mesenchymal stem cell differentiation. Methods Mol Biol. 2011;698:459–84.
100. Hang HC, et al. Bioorthogonal chemical reporters for analyzing protein lipidation and lipid trafficking. Acc Chem Res. 2011;44(9):699–708.
101. Heal WP, et al. N-Myristoyl transferase-mediated protein labelling in vivo. Org Biomol Chem. 2008;6(13):2308–15.
102. Kramer G, et al. Identification and quantitation of newly synthesized proteins in Escherichia coli by enrichment of azidohomoalanine-labeled peptides with diagonal chromatography. Mol Cell Proteomics. 2009;8(7):1599–611.
103. Kramer G, et al. Proteome-wide alterations in Escherichia coli translation rates upon anaerobiosis. Mol Cell Proteomics. 2010;9(11):2508–16.
104. Li Y, et al. Removal of a single N-linked glycan in human immunodeficiency virus type 1 gp120 results in an enhanced ability to induce neutralizing antibody responses. J Virol. 2008;82(2):638–51.
105. Reitter JN, et al. A role for carbohydrates in immune evasion in AIDS. Nat Med. 1998;4(6):679–84.
106. Huskens D, et al. The role of N-glycosylation sites on the CXCR4 receptor for CXCL-12 binding and signaling and X4 HIV-1 viral infectivity. Virology. 2007;363(2):280–7.
107. Linde ME, et al. The conserved set of host proteins incorporated into HIV-1 virions suggests a common egress pathway in multiple cell types. J Proteome Res. 2013;12(5):2045–54.
108. Angel TE, et al. The cerebrospinal fluid proteome in HIV infection: change associated with disease severity. Clin Proteomics. 2012;9(1):3.
109. Berro R, et al. Identifying the membrane proteome of HIV-1 latently infected cells. J Biol Chem. 2007;282(11):8207–18.
110. Kadiu I, et al. HIV-1 transforms the monocyte plasma membrane proteome. Cell Immunol. 2009;258(1):44–58.
111. Chertova E, et al. Proteomic and biochemical analysis of purified human immunodeficiency virus type 1 produced from infected monocyte-derived macrophages. J Virol. 2006;80(18):9039–52.
112. Alfaro JF, et al. Tandem mass spectrometry identifies many mouse brain O-GlcNAcylated proteins including EGF domain-specific O-GlcNAc transferase targets. Proc Natl Acad Sci U S A. 2012;109(19):7280–5.
113. Nagaraj N, et al. Correction to feasibility of large-scale phosphoproteomics with higher energy collisional dissociation fragmentation. J Proteome Res. 2012;11(6):3506–8.
114. Hanisch FG. O-glycoproteomics: site-specific O-glycoprotein analysis by CID/ETD electrospray ionization tandem mass spectrometry and top-down glycoprotein sequencing by in-source decay MALDI mass spectrometry. Methods Mol Biol. 2012;842:179–89.

115. Binley JM, et al. Role of complex carbohydrates in human immunodeficiency virus type 1 infection and resistance to antibody neutralization. J Virol. 2010;84(11):5637–55.
116. Gouveia R, et al. Expression of glycogenes in differentiating human NT2N neurons. Downregulation of fucosyltransferase 9 leads to decreased Lewis(x) levels and impaired neurite outgrowth. Biochim Biophys Acta. 2012;1820(12):2007–19.
117. Takahashi N. Demonstration of a new amidase acting on glycopeptides. Biochem Biophys Res Commun. 1977;76(4):1194–201.
118. Yang SJ, Zhang H. Glycan analysis by reversible reaction to hydrazide beads and mass spectrometry. Anal Chem. 2012;84(5):2232–8.
119. Nwosu CC, et al. Comparison of the human and bovine milk N-glycome via high-performance microfluidic chip liquid chromatography and tandem mass spectrometry. J Proteome Res. 2012;11(5):2912–24.
120. Froehlich JW, et al. Nano-LC-MS/MS of glycopeptides produced by nonspecific proteolysis enables rapid and extensive site-specific glycosylation determination. Anal Chem. 2011;83(14):5541–7.
121. Hua S, et al. Site-specific protein glycosylation analysis with glycan isomer differentiation. Anal Bioanal Chem. 2012;403(5):1291–302.
122. Krishnamoorthy L, et al. HIV-1 and microvesicles from T cells share a common glycome, arguing for a common origin. Nat Chem Biol. 2009;5(4):244–50.
123. Yang J, Caprioli RM. Matrix sublimation/recrystallization for imaging proteins by mass spectrometry at high spatial resolution. Anal Chem. 2011;83(14):5728–34.
124. Bruand J, et al. Automated querying and identification of novel peptides using MALDI mass spectrometric imaging. J Proteome Res. 2011;10(4):1915–28.
125. Goodwin RJ, et al. Protein and peptides in pictures: imaging with MALDI mass spectrometry. Proteomics. 2008;8(18):3785–800.
126. Seeley EH, Caprioli RM. 3D imaging by mass spectrometry: a new frontier. Anal Chem. 2012;84(5):2105–10.
127. Robinson S, et al. Localization of water-soluble carbohydrates in wheat stems using imaging matrix-assisted laser desorption ionization mass spectrometry. New Phytol. 2007;173(2):438–44.
128. Ott M, et al. Tat acetylation: a regulatory switch between early and late phases in HIV transcription elongation. Novartis Found Symp. 2004;259:182–93; discussion 186–193, 185–223.
129. Raska M, et al. Glycosylation patterns of HIV-1 gp120 depend on the type of expressing cells and affect antibody recognition. J Biol Chem. 2010;285(27):20860–9.
130. Bentham M, et al. Role of myristoylation and N-terminal basic residues in membrane association of the human immunodeficiency virus type 1 Nef protein. J Gen Virol. 2006;87(Pt 3):563–71.
131. Rousso I, et al. Palmitoylation of the HIV-1 envelope glycoprotein is critical for viral infectivity. Proc Natl Acad Sci U S A. 2000;97(25):13523–5.
132. Strack B, et al. A role for ubiquitin ligase recruitment in retrovirus release. Proc Natl Acad Sci U S A. 2000;97(24):13063–8.

Chapter 7
Bioinformatics for Mass Spectrometry-Based Proteomics

Rebekah L. Gundry

Introduction

Chapter 4 focused on sample preparation and fundamentals regarding design and execution of a successful mass spectrometry (MS) experiment to obtain high-quality data. In this chapter, readers are introduced to the processes used to convert a set of MS data into an annotated dataset of confidently identified and characterized proteins in the context of studies involving HIV-1, particularly in the management and structure of databases. Fundamental MS data types are reviewed followed by a description of how to select an appropriate protein database and use it within a search. Subsequently, important concepts regarding data organization, visualization, and biological information mining are discussed. Finally, the chapter concludes with a view on the future of systems biology studies using HIV-1 proteomics. As this chapter focuses on introducing the reader to the most commonly encountered bioinformatic concepts in modern HIV-1 proteomics, this discussion is focused on data produced by untargeted, bottom-up approaches, whereby the protein is first chemically or enzymatically digested into peptides prior to MS analysis, and the goal is to collect data for as many analytes as possible without prior knowledge of the proteins present in the sample. However, readers should be aware of other powerful strategies that will require different bioinformatics approaches than discussed here. These include *top-down proteomics*, whereby intact proteins are fragmented during MS analysis to obtain amino acid sequence information [1, 2] in a way that preserves the stoichiometry among posttranslational modifications, proteolytic cleavage products, and products of splicing events, among others; *targeted proteomics*, where preselected peptides or proteins of interest are quantified and/or

R.L. Gundry, Ph.D. (✉)
Department of Biochemistry, Medical College of Wisconsin,
8701 Watertown Plank Road, Milwaukee, WI 53226, USA
e-mail: rgundry@mcw.edu

D.R.M. Graham, D.E. Ott (eds.), *HIV-1 Proteomics*,
DOI 10.1007/978-1-4939-6542-7_7

characterized; and *data-independent acquisition* where peptides are selected for fragmentation independent of their signal intensity [3]. Overall, these approaches are often complementary to the data-dependent, bottom-up proteomic strategies which are the focus of this chapter. Ultimately, HIV-1 biology will benefit from each of these approaches which collectively can be used to identify, quantify, and characterize proteins and peptides from cells, tissues, and biological fluids.

Fundamentals of Peptide MS Data

Spectra, Chromatograms, and Ion Maps: When performing liquid chromatography MS (LC-MS), peptides are eluted from the chromatographic column over time and directly introduced to the mass spectrometer. The MS instrumentation will record two main types of spectra during a bottom-up data-dependent acquisition. A *precursor ion spectrum or scan*, also referred to as an MS^1 or MS spectrum, records a snapshot of mass-to-charge (*m/z*) values corresponding to peptide masses present at a particular time in the elution profile. A *product ion spectrum or scan*, also referred to as an MS^2 or MS/MS spectrum, records *m/z* values corresponding to ions resulting from the fragmentation of a peptide ion selected from the MS spectrum. In a single LC-MS experiment, the instrument is continually cycling between recording MS and MS/MS spectra. LC-MS data can be viewed as a chromatogram composed of a large set of consecutively acquired mass spectra. It is possible to view these data as a *base peak chromatogram* where the signal of the ions giving the base peak (i.e., the most intense peak in each spectrum) in each of a series of spectra is plotted as a function of elution time. Important for quantitation, it is possible to view an *extracted ion chromatogram*, where the intensity of a selected *m/z* value or set of values are plotted as a function of elution time. In cases where chromatographic performance of the system is excellent and highly reproducible, the area under the curve for the extracted ion chromatogram of a particular peptide can be used for quantitation. LC-MS data can also be displayed as a three-dimensional ion map where each point represents the intensity for a particular *m/z* value at a particular elution time, which can be used in label-free quantitation workflows. The relationship of an ion map, precursor ion spectrum, and product ion spectrum is shown in Fig. 7.1.

Peptide Fragmentation: To understand MS/MS spectra and how they are used in a proteomic workflow to identify proteins, it is important to have a basic understanding of the fragmentation process that occurs inside the MS instrument (see Chap. 3). Briefly, in an untargeted peptide-centric mass spectrometry experiment, the instrument isolates one ion representative of a peptide and subjects it to fragmentation. Peptide fragmentation can be carried out by a variety of means, typically by collision with an inert gas or radical anions. While the masses resulting from this fragmentation process will vary depending on the fragmentation type used, the resulting masses (i.e., fragment ions) are directly related to the amino acid sequence of the peptide. Therefore, once high-quality MS/MS data are recorded, modern bioinformatics approaches can use this information to rapidly determine protein identity and sites of posttranslational modifications.

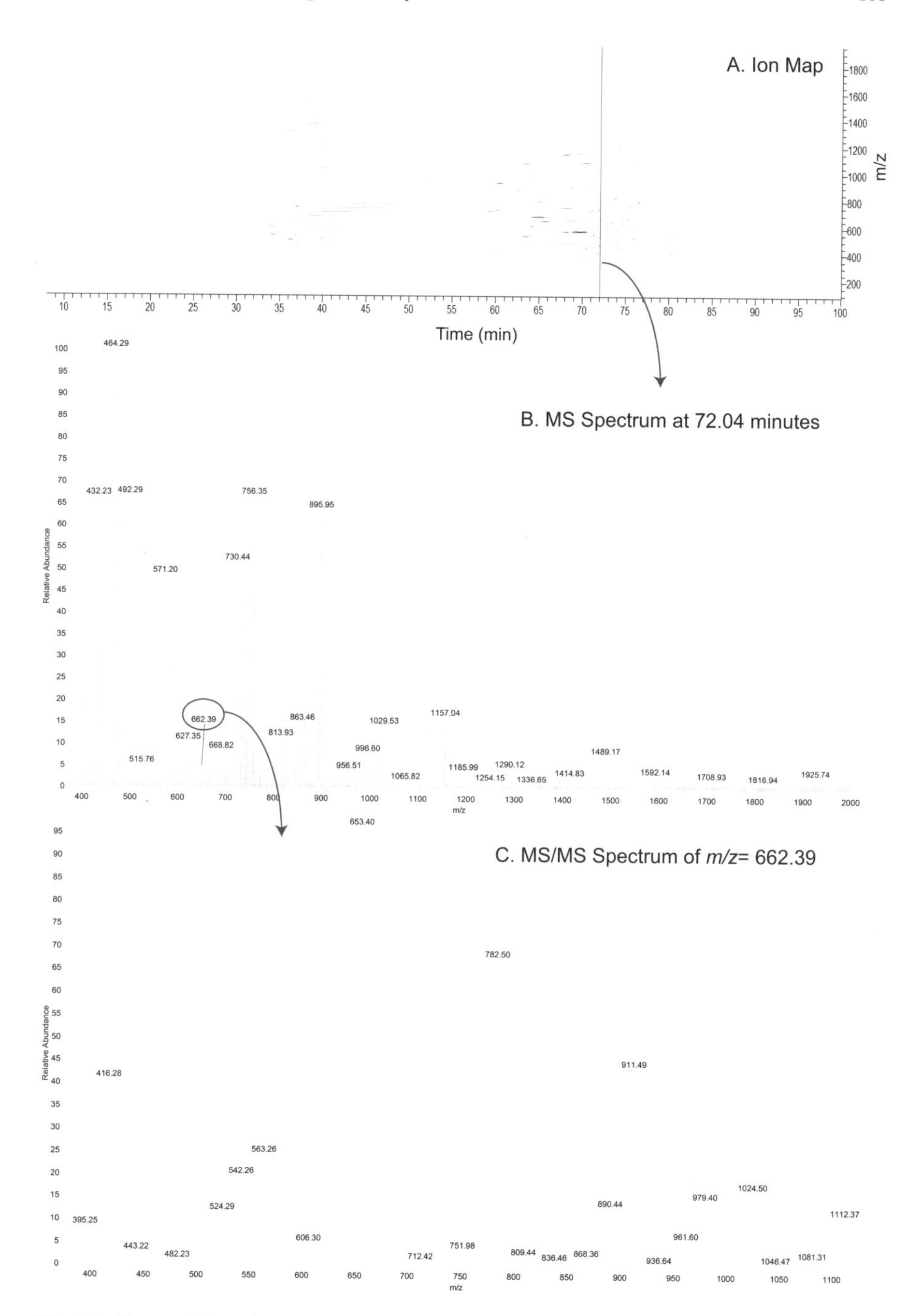

Fig. 7.1 Types of data obtained during an LC-MS/MS proteomics experiment. (**a**) A representative three-dimensional ion map where intensity values (color density) for each *m/z* observed over time are represented. (**b**) Precursor ion spectrum for 72.04 min. (**c**) Product spectrum for *m/z*=662.39. Data were obtained on an LTQ Orbitrap (Thermo) and viewed using Xcalibur™ software (Thermo)

Fundamentals of the Proteomic Workflow

The proteomic workflow conceptually includes all steps involved in the process of converting a set of MS and MS/MS spectra into an annotated dataset of confidently identified, characterized, and, sometimes, quantified proteins (Fig. 7.2). This process includes peak extraction, searching against a relevant protein sequence database or de novo sequencing, post-search validation and redundancy removal, and ends with augmentation of the dataset with annotations or other known or predicted functional information. At the time of writing this edition, the reader can benefit

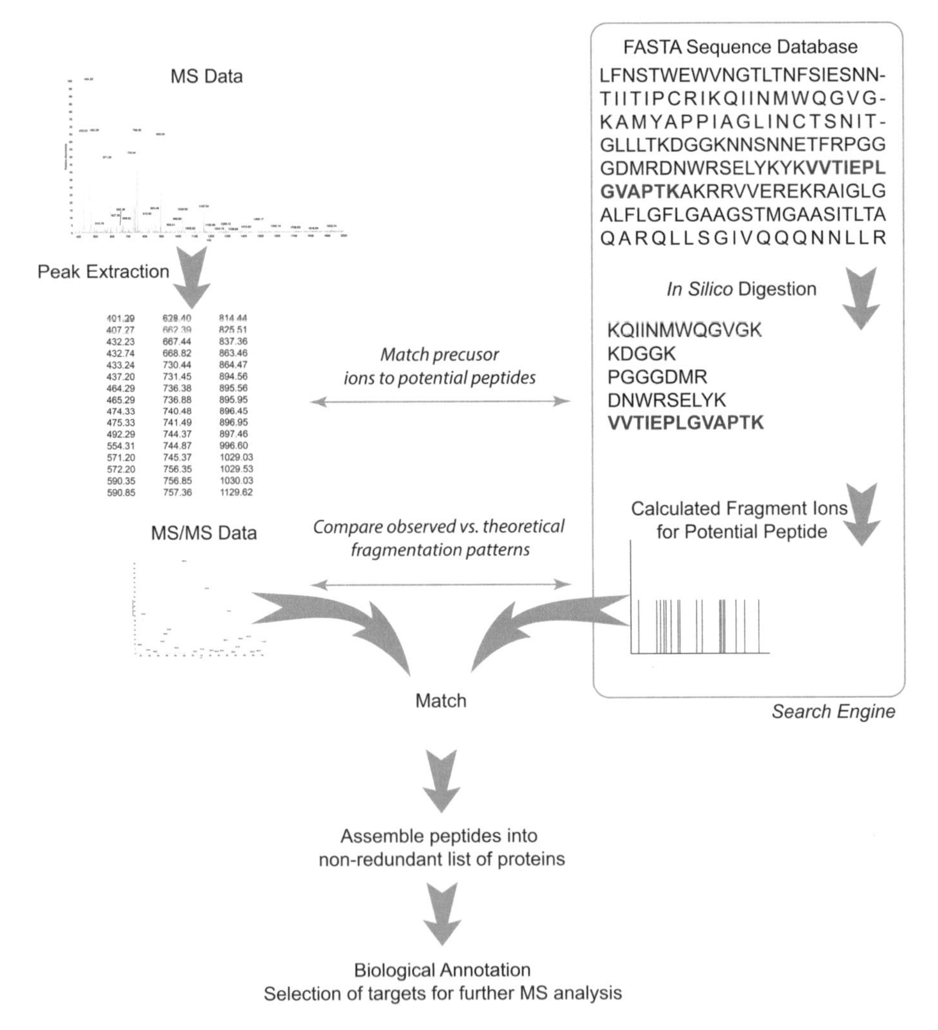

Fig. 7.2 Simplified overview of the bioinformatics workflow for processing MS data from an untargeted, bottom-up proteomics experiment

from over a decade in the evolution of bioinformatic tools for proteomic data which has yielded many commercial and open-source tools for each step of the process. As space does not permit an in-depth discussion of all tools available, readers are directed to several excellent reviews and internet resources that comprehensively summarize the tools currently available for each aspect of the proteomic workflow. Ultimately, the choice of preferred platform depends on budget, instrumentation used for data acquisition, biological questions, and user expertise.

MS Peak Extraction: The detection and extraction of peaks plays a critical role in the proteomic workflow. Reliable peptide and protein identification requires accurate and precise determination of the *m/z* values of the ions. Challenges to peak extraction can include chemical noise, baseline artifacts, asymmetrical peak shape, low mass resolution, and high charge stages. While some of these challenges can be addressed through the use of modern instrumentation with high resolving power and mass accuracy, there remains the need for high-performance bioinformatics approaches to maximize the quantity and quality of extracted data. Most vendors provide peak extraction software with the instrumentation, and some database search platforms include supplementary peak extraction tools. As a result, there are several open-source and commercial tools that can be used to improve peak extraction efficiency, and Mancuso and colleagues recently demonstrated the benefit of using optimized peak extraction algorithms for high mass accuracy and resolution data recorded using an orbitrap by comparing performance of nine different extraction tools [4]. While some of the commercial peak picking tools are proprietary, there has been a commitment by instrumentation vendors and bioinformatics developers to work closely together to ensure that MS data from most modern instrumentation can be processed by any analysis platform, either directly or by using industry standard output format mzML [5]. This process is aided by a variety of freely available convertors that turn vendor-specific data into the searchable mzML format, and many of these tools are summarized at expasy.org, tools.proteomecenter.org, proteinscience.com, and ms-utils.org.

Selecting Appropriate Protein Sequence Databases: In the most common proteomic workflow, an automated search engine is used to assign peptide sequences to the MS/MS spectra. This process requires a protein sequence database to serve as the source of potential peptide sequences that can be matched to the experimental data. Search algorithms will typically require that the database to be searched against be provided as a FASTA formatted file, which is a simple text file of protein sequences represented using single-letter codes. Figure 7.3 shows an example of a FASTA sequence for an HIV-1 envelope glycoprotein.

While numerous web resources for FASTA files are available, the HIV researcher should carefully consider the source and content appropriate for the study. Of course, if a peptide that gave rise to the MS/MS spectrum is not present in the sequence database used to process the data, the search algorithm will not be able to match the MS/MS spectrum to the correct peptide sequence. The National Center for Biotechnology Information (ncbi.nlm.nih.gov) and other public repositories like the Swiss Institute of Bioinformatics (isb-sib.ch) and the European Bioinformatics Institute (part of the European Molecular Biology Laboratory—ebi.ac.uk) have

>gi|973446878|gb|ALX35231.1| envelope glycoprotein [Human immunodeficiency virus 1]
MRVRGTGRNYQHLWKWGTMLLGLLLMTYSTAEDSWVTVYYGVPVWKEAKAILFCASDAKAYETEAHNIWA
THACVPTDPNPQEIKLENVTENFNMWKNNMVEQMHEDIISLWDESLKPCVKLTPLCVTLNCSDVKVNNTS
TNAATNSSSNSTAAPMTTVASSSDMKNCSFNITTAVRDEQKKVHSLFYKLDIVNMEGKNNTGDGTYRLIN
CNTTTITQACPKVSFEPIPIHYCAPAGFAILKCNDKQFNGTGPCTNVSTVQCTHGIRPVVSTQLLLNGSL
AEEKIMIRSENLTSNSKTIIVQLNESLSITCIRPNNNTRQGVGIGPGQMFFTTGIIGDIRQAHCNISGTK
WNSTLQKVAEELKDRLNITKIIFKPHSGGDPEITTHSFNCGGEFFYCNTSKLFNSTWEWVNGTLTNFSIE
SNNTIITIPCRIKQIINMWQGVGKAMYAPPIAGLINCTSNITGLLLTKDGGKNNSNNETFRPGGGDMRDN
WRSELYKYK**VVTIEPLGVAPTK**AKRRVVEREKRAIGLGALFLGFLGAAGSTMGAASITLTAQARQLLSGI
VQQQNNLLRAIEAQQHMLQLTVWGIKQLQARVLAVERYLKDQQLLGIWGCSGKHICTTSVPWNSTWSNKT
LEEIWQNMTWMQWEREIDNYTDEIYNLLADSQLQQEHNEKELLELDKWASLWNWFDISQWLWYIKIFIMI
VGGLIGLRIFFAVLSIVNRVRQGYSPLSFQTLLPAPRGPDRPEGIEEEGGDNDRGRSTRLVNGLSALIWD
DLRNLCLFSYRRLRDLVLIAARTVELLGRRGWETLKYLWNLLQYWSQELKNSAVSLLNTTAIAVAEGTDR
IIEVVQRACRAVLHIPRRIRQGFERALL

Fig. 7.3 A representative FASTA sequence of an HIV-1 envelope glycoprotein. The amino acids in *red* correspond to a peptide whose annotated MS/MS spectrum is shown in Fig. 7.4

been collecting and annotating protein sequences into a variety of databases. More recently, there have been highly collaborative efforts between these partners to form the UniProt consortium (uniprot.org). For many years, the UniProtKB/Swiss-Prot database has been the source of choice for most organisms, which at the time of writing contains 548,546 entries comprising 195,452,300 amino acids abstracted from 236,982 references (source: expasy.org, release 2015_06 statistics). By far, UniProtKB/Swiss-Prot is the most common and arguably the top reference database for use in human proteomic studies. However, while UniProtKB/Swiss-Prot contains 20,206 entries for human, it has <2000 entries for macaques of different origins and only 9 entries for HIV-1. When extending out to other UniProt resources like TrEMBL, which is unreviewed, there are >125,000 entries for humans, ~70,000 entries for *Macaca mulatta*, but only ~3700 sequence entries for HIV-1. Fortunately, since 1987, the Theoretical Biology and Biophysics Group at the Los Alamos National Laboratory (LANL) has been funded by the Division of AIDS of the National Institute of Allergy and Infectious Diseases to accept and curate HIV-1 sequences. These sequences, and a variety of tools to perform HIV-1 bioinformatics, are freely available (hiv.lanl.gov). Most of the sequences found in LANL are derived from GenBank, which at the time of writing includes over 765,000 entries for HIV. A guide to the HIV sequence database and the other major HIV database (the HIV RT/Protease Sequence Database at Stanford) was published in 2003 by Kuiken and colleagues and provides useful information on how to use these resources effectively [6]. Ultimately, investigators must exercise caution when selecting a FASTA database to use in a proteomic analysis, as the number and extent to which the sequences have been manually curated will vary.

Given the low costs of generating HIV and SIV sequence information as described in Chap. 4, generating an experiment-specific database is now practical for scientists wishing to perform HIV-1 proteomics experiments. If limited sequence information is available for the virus of interest, it is possible to generate additional sequence information by using BLASTP to find virus sequences similar to the target sequence. This type of in silico inflation can be a useful tool to increase the sequence variability around a specific known sequence within a FASTA database, without overburdening the bioinformatic analysis (e.g., if all HIV sequences were included). Until more recently, appending all of these redundant sequences belonging to different strains of HIV could have created a challenge through increasing the redundancy present in the database; however, the use of alignment and clustering algorithms in modern bioinformatics platforms largely overcomes these challenges (discussed below).

If merging multiple FASTA files is required, a text editor capable of editing large files (e.g., www.vim.org; macupdate.com) is adequate. However, the novice user can quickly become lost or unknowingly generate errors. To avoid potential errors due to manual manipulation, users should consider computational resources that eliminate manual intervention and manipulate FASTA databases quickly and efficiently, including dbtoolkit and fastahack available at github.com and subsetDB at tools.proteomecenter.org. Moreover, in recent versions of Proteome Discoverer (Thermo), it is possible to include one or many separate FASTA files within a single search without merging them in advance. Ultimately, the investigator should become familiar with the benefits and methods for creating custom FASTA databases for proteomics research. Investigators new to this approach may find it useful to review the benefits of using carefully curated databases in a "fit-for-purpose" approach [7], as well as visit the Computational Omics and Systems Biology Group (compomics.com) which provides a rich resource for educational materials describing the principles of database searching and how database size and accuracy can affect search results. Specifically, the workshop on MS data processing illustrates the problem of information inflation that can occur when redundant entries are added to databases without high information quality (http://compomics.com/workshops/bits-ms-data-processing/).

The Database Search: The fundamental goal of the search algorithm (or search engine) is to match an MS/MS spectrum obtained during the experiment to a peptide sequence from a predefined database (i.e., to generate peptide spectrum matches). The output is a list of peptide sequences that can explain the experimental data with a certain degree of probability or false discovery rate. As described above, the databases are typically protein sequences translated from genomic data, although spectral libraries [8] or mRNA data [9] can also be used. These identified peptides are then assembled into proteins, which is where it can become computationally challenging if the database contains redundant peptides or alternatively spliced proteins.

As mentioned above, many algorithms have been developed for searching peptide mass spectrometry data; however, there is considerable variation in the level of in-house bioinformatics expertise and support required to install and use these tools properly. Furthermore, the options that each platform provides for enhanced data visualization can vary significantly. This is especially important in the case of HIV-1 proteomics, where there can literally be thousands of matches for the Gag polyprotein, with annotations belonging to thousands of different entries in the database. This can make data interpretation exceptionally challenging. For this reason, easy-to-use platforms with enhanced data visualization capabilities can be especially beneficial. While it is not feasible to comprehensively review all of the available software tools in this chapter, several examples are provided and readers are directed to several recent reviews and Wikipedia https://en.wikipedia.org/wiki/List_of_mass_spectrometry_software) that have comprehensively summarized available platforms [10–12].

Of the freely available tools, those at CompOmics (www.compomics.com) are easy to install and capable of performing database searching and data visualization. These tools include dbtoolkit, described above, and SearchGUI [13], which is a graphical user interface for running proteomics search engines including X! Tandem [13–19], MS-GF+ [20], MS Amanda [21], MyriMatch [22], Comet [23], Tide [24], Andromeda [25], and OMSSA [13, 26–29]. Finally, PeptideShaker [30] is a free alternative to Scaffold (Proteome Software) and allows for the visualization of a proteomics experiment generated from SearchGUI using the tools described above and supports the full export of data to the PRIDE database, one of the leading public repositories for proteomic data. Furthermore, PeptideShaker can be used to import data from the PRIDE database, thus allowing collaborators to view data without having to distribute files or install the same software platforms. A commercial alternative to PeptideShaker is Scaffold software, a post-search processing tool that is used to visualize search results from multiple search engines including Mascot [31], Mascot Distiller (Matrix Science), Proteome Discoverer (Thermo Scientific), Spectrum Mill (Agilent), SEQUEST [32], Phenyx (includes OLAV scoring [33]), IdentityE/PLGS (Waters Corporation), X! Tandem [13–19], OMSSA [26], and MaxQuant/Andromeda [25]. The benefits of Scaffold is that it automates FDR scoring, allows for protein and peptide FDR filtering and family clustering, and includes validation tools like PeptideProphet/ProteinProphet [34] and additional statistical features. While the investment for academic laboratories is modest, the plug-n-play functionality and data visualization interface can enable alignment of matches to a consensus HIV-1 sequence, allowing quick interpretation of data. Another commercial software tool that is especially relevant for HIV-1 proteomics is PEAKS [35]. PEAKS enables traditional database searching approaches using its own algorithm and can incorporate multiple algorithm searching using Mascot, OMSSA, SEQUEST, and X! Tandem (ibid). Moreover, PEAKS includes de novo sequencing tools and sequence tag homology searching and integrates a posttranslational modification finder, each of which can be especially useful for studying organisms where protein databases may be incomplete. PEAKS can be used to search spectra that have gone unmatched in other searches and thereby has the potential to increase protein coverage of HIV-1 proteins and identify novel posttranslational modifications.

Independent of the search algorithm and post-search validation tools used, the availability of spectral annotation viewers allows investigators to quickly visually inspect the quality of the MS/MS spectra. In this process, particularly when amino acid variants or posttranslational modifications are suspected, investigators will ensure that the majority of peaks are assigned to the predicted peptide sequence and, vice versa, that a core set of fragment ions predicted from the proposed sequence are observed in the spectrum (e.g., series of y-ions). The outcome of the proteomic workflow is exemplified in Fig. 7.4. Figure 7.4a reviews the major types of ions that are observed when various fragmentation approaches are used. Figure 7.4b shows *m/z* values for fragment ions that are calculated to result from CID fragmentation of a proposed peptide sequence for the MS/MS spectrum obtained in Fig. 7.1c, and those peaks observed in the MS/MS spectrum are highlighted in color. Figure 7.4c shows the annotated MS/MS spectrum of precursor $m/z = 662.39$, where all major

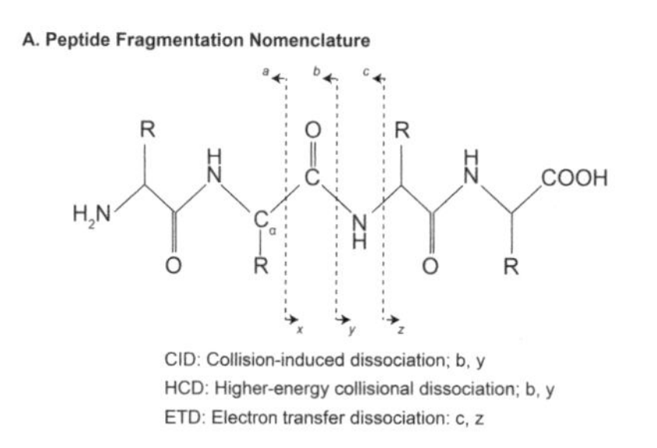

B. Predicted fragment ions from peptide VVTIEPLGVAPTK

#	b⁺	b²⁺	Sequence	y⁺	y²⁺	#
1	100.07569	50.54148	V			13
2	199.14410	100.07569	V	1224.71981	612.86355	12
3	300.19178	150.59953	T	1125.65140	563.32934	11
4	413.27585	207.14156	I	1024.60372	512.80550	10
5	542.31844	271.66286	E	911.51966	456.26347	9
6	639.37120	320.18924	P	782.47707	391.74217	8
7	752.45527	376.73127	L	685.42430	343.21579	7
8	809.47673	405.24200	G	572.34024	286.67376	6
9	908.54515	454.77621	V	515.31877	258.16303	5
10	979.58226	490.29477	A	416.25036	208.62882	4
11	1076.63502	538.82115	P	345.21325	173.11026	3
12	1177.68270	589.34499	T	248.16048	124.58388	2
13			K	147.11280	74.06004	1

C. Annotated MS/MS Spectrum

Peptide assignment: VVTIEPLGVAPTK from HIV-1 Envelope glycoprotein
Precursor m/z = 662.39
Precursor Mass Error: 0.76 ppm

Fig. 7.4 Peptide fragmentation nomenclature and how it is used to interpret and annotate an MS/MS spectrum. (**a**) Peptide fragmentation nomenclature, indicating the types of fragment ions most commonly observed for each of the three major fragmentation mechanisms used in modern proteomics. For a particular cleavage position, the assignment of *a*, *b*, *c* vs. *x*, *y*, *z* is determined by the terminus (N- vs. C-terminus) on which the charge remains (indicated by *arrows*). (**b**) *m/z* values for fragment ions predicted to result from the proposed peptide sequence VVTIEPLGVAPTK from HIV-1 envelope glycoprotein. Values in *red* or *blue* were observed in the MS/MS spectrum in panel **c**. (**c**) Annotated MS/MS spectrum from CID fragmentation of precursor *m/z* 662.39 acquired using an LTQ Orbitrap MS (raw data shown in Fig. 7.1c). Assigned fragment ions are indicated and correspond to predictions in panel **b**. Panels **b** and **c** were generated using Proteome Discoverer 2.1 (Thermo)

peaks are accounted for by the proposed peptide sequence from HIV-1 envelope glycoprotein. In summary, by following the example of HIV-1 envelope glycoprotein outlined in this chapter, the reader is introduced to raw MS data (Fig. 7.1), the proteomic workflow (Fig. 7.2) that is used to process the raw data using a FASTA sequence (Fig. 7.3), and the outcome of the data interpretation (Fig. 7.4).

Posttranslational Modifications: Chapter 6 discussed strategies for studying various posttranslational modifications involved in the HIV-1 life cycle. This topic is worth revisiting in the context of informatics, as the appropriate scope and number of modifications to allow during a search will impact results. For example, including too many potential posttranslational modifications in the search can artificially inflate the number of potential matches for each spectrum. Typically, the minimalistic set of modifications that are most frequently associated with the specific experimental conditions will yield the most robust results, and for this reason, it is important to consider stoichiometry for each potential modification. For example, reduction and alkylation are common steps to encourage protein unfolding and thereby enhance enzymatic digestion. If this is used during sample preparation, then a modification (e.g., carbamidomethylation) at cysteine residues is allowed in the search. Similarly, oxidation at methionine and N-terminal protein acetylation are also commonly included. Beyond these common modifications, the choice to include additional modifications depends on stoichiometry and whether it is logical that a modification that occurs in low abundance should appear as a unique match. In the case of phosphorylation, it is typically unlikely to routinely observe this modification unless the sample was specifically prepared to enrich for phosphorylated peptides. Therefore, if a phosphopeptide enrichment is not performed, including phosphorylation in the first round of searching may lead to increased rate of false-positive identifications. Bioinformatics approaches that include additional rounds of searching [36, 37] for spectra that go unmatched in the first round can be powerful for assigning high-confidence posttranslational modifications even without physical enrichment. Moreover, Scaffold PTM, PEAKS, Byonic [38], and PTMrs [39], for example, can improve site localization assignment of a wide variety of posttranslational modifications.

The Redundancy and Annotation Problem: In the context of HIV-1 proteomics, each polymorphism can result in a separate FASTA entry in the database and then result in thousands of different "hits" for different viruses and often hundreds of hits for single viral proteins. Therefore, the raw output of a single search algorithm can result in an unmanageable number of matches. Fortunately, many software packages like Scaffold, Proteome Discoverer, and ProteinCenter (Thermo) have very good approaches to rapidly cluster data into meaningful outputs.

Extracting Annotated and Predicted Biology

Once a set of peptide and protein identifications have been made, the next step in the proteomic workflow is dependent upon the biological question or experimental goals. In some cases, after the first round of discovery, it may be apt to reanalyze the samples using a targeted approach, where quantitative measurements of preselected peptides can be made with more precision and lower detection limits than possible in the non-targeted discovery approach. This strategy can, for example, provide further confidence in putative abundance changes measured in the first round of discovery or be used to obtain more evidence for a novel posttranslational modification. In the case of HIV-1, targeted analysis of specific peptides that are signatures of intact vs. processed gag can inform the degree of virus maturation. Also, this approach can be used to compare the relative abundance of the envelope glycoprotein among samples. Alternatively, the next step may be additional bioinformatic analyses of the proteins identified in the discovery experiment. In this case, when working with organisms whose protein sequence databases are only partially or recently completed, the descriptions for each entry may be incomplete. Hence, the proteins may lack information on biological function, subcellular localization, interactions, and pathway affiliations. Therefore, in an effort to begin to understand a particular protein's biological role, it can be helpful to analyze the amino acid sequence for similarity to other proteins that are annotated, for specific folds of protein domains, and for sequence motifs that may indicate propensity for a particular posttranslational modification. A BLAST (basic local alignment search tool, NCBI) search against a database of known protein sequences can be used to identify proteins with similar amino acid sequences that have been described in the same or other organisms (http://blast.ncbi.nlm.nih.gov). Numerous tools to determine if the identified protein shares a specific protein fold or domain with other characterized proteins can be found on InterPro [40] (http://www.ebi.ac.uk/interpro/), thereby potentially helping to associate the unknown protein with a family to which it is most closely related. Finally, more than 30 tools are summarized at http://www.expasy.org/proteomics that can be used to scan the amino acid sequence for motifs that are commonly associated with particular posttranslational modifications, thereby predicting the probability that a protein contains a particular modification and, in some cases, predicting which modifying enzyme (e.g., a particular kinase) is most likely to act on that site. Beyond the abovementioned resources, interactive views of structure, function, interactions, and genomic location of HIV-1 proteins can be found at the BioAfrica HIV-1 Proteomics Resource [41] (http://www.bioafrica.net/proteomics/).

The Future of HIV Proteomics

Overall, the integration of modern proteomic sample preparation techniques, mass spectrometry, and bioinformatics are poised to have a tremendous impact on HIV biology. Beyond the study of virions, the principles described here can be applied to larger-scale systems biology studies of HIV-1, as shown in recent examples that investigated protein–protein interactions critical to HIV replication [42] and glycoproteins associated with latent infection [43]. By applying some of the approaches described here for obtaining additional resolution and sequence coverage of viral proteins, HIV-1 proteomics approaches can be applied to determine the relative abundance of various viral polymorphisms and relative abundance of mutations in regions where amplification approaches may be insufficient or for identifying a specific region or posttranslational modification of a region of a viral protein that is responsible for host–viral interactions. Undoubtedly, it is exciting to envision how mass spectrometry could be used to detect virus during primary infection in ways that rival the sensitivity of current amplification-based clinical assays, to characterize viral envelope proteins in a way that explains how HIV-1 evades destruction by immune cells, or to determine whether viral proteins or the virus itself directly is responsible for CD4 T-cell toxicity. While current evidence suggests a promising outlook, technical challenges must be acknowledged and addressed. Experience has taught us to be mindful of the fundamental details within each discipline that can make or break an analysis. For all these reasons, synergistic efforts among scientists with expertise in HIV biology, protein chemistry, mass spectrometry, bioinformatics, and systems biology that align with their clinical counterparts are imperative to embarking on truly innovative studies for expanding our understanding of HIV biology and developing advanced clinical therapies.

Acknowledgments This work was supported by the National Institutes of Health (R01HL126785 and R01HL134010). Special thanks to Dr. David Graham (Johns Hopkins University) for providing raw MS data used to generate Figs. 7.1 and 7.4 and Dr. Claudius Mahr (University of Washington) for editorial assistance.

References

1. Peng Y, Ayaz-Guner S, Yu D, Ge Y. Top-down mass spectrometry of cardiac myofilament proteins in health and disease. Proteomics Clin Appl. 2014;8(7–8):554–68.
2. Savaryn JP, Catherman AD, Thomas PM, Abecassis MM, Kelleher NL. The emergence of top-down proteomics in clinical research. Genome Med. 2013;5(6):53.
3. Venable JD, Dong MQ, Wohlschlegel J, Dillin A, Yates JR. Automated approach for quantitative analysis of complex peptide mixtures from tandem mass spectra. Nat Methods. 2004;1(1):39–45.
4. Mancuso F, Bunkenborg J, Wierer M, Molina H. Data extraction from proteomics raw data: an evaluation of nine tandem MS tools using a large Orbitrap data set. J Proteomics. 2012;75(17):5293–303.
5. Martens L, Chambers M, Sturm M, Kessner D, Levander F, Shofstahl J, et al. mzML—a community standard for mass spectrometry data. Mol Cell Proteomics. 2011;10(1):R110.000133.
6. Kuiken C, Korber B, Shafer RW. HIV sequence databases. AIDS Rev. 2003;5(1):52–61.
7. Cheng K, Sloan A, McCorrister S, Babiuk S, Bowden TR, Wang G, et al. Fit-for-purpose curated database application in mass spectrometry-based targeted protein identification and validation. BMC Res Notes. 2014;7:444.
8. Lam H. Building and searching tandem mass spectral libraries for peptide identification. Mol Cell Proteomics. 2011;10(12):R111.008565.
9. Wang X, Slebos RJ, Wang D, Halvey PJ, Tabb DL, Liebler DC, et al. Protein identification using customized protein sequence databases derived from RNA-Seq data. J Proteome Res. 2012;11(2):1009–17.
10. Cottrell JS. Protein identification using MS/MS data. J Proteomics. 2011;74(10):1842–51.
11. Eng JK, Searle BC, Clauser KR, Tabb DL. A face in the crowd: recognizing peptides through database search. Mol Cell Proteomics. 2011;10(11):R111.009522.
12. Kapp E, Schutz F. Overview of tandem mass spectrometry (MS/MS) database search algorithms. Curr Protoc Protein Sci 2007;Chapter 25:Unit 25.2
13. Vaudel M, Barsnes H, Berven FS, Sickmann A, Martens L. SearchGUI: an open-source graphical user interface for simultaneous OMSSA and X!Tandem searches. Proteomics. 2011;11(5):996–9.
14. Yang P, Ma J, Wang P, Zhu Y, Zhou BB, Yang YH. Improving X!Tandem on peptide identification from mass spectrometry by self-boosted percolator. IEEE/ACM Trans Comput Biol Bioinform. 2012;9(5):1273–80.
15. Pratt B, Howbert JJ, Tasman NI, Nilsson EJ. MR-Tandem: parallel X!Tandem using Hadoop MapReduce on Amazon Web Services. Bioinformatics. 2012;28(1):136–7.
16. Muth T, Vaudel M, Barsnes H, Martens L, Sickmann A. XTandem Parser: an open-source library to parse and analyse X!Tandem MS/MS search results. Proteomics. 2010;10(7):1522–4.
17. Brosch M, Swamy S, Hubbard T, Choudhary J. Comparison of Mascot and X!Tandem performance for low and high accuracy mass spectrometry and the development of an adjusted Mascot threshold. Mol Cell Proteomics. 2008;7(5):962–70.
18. Bjornson RD, Carriero NJ, Colangelo C, Shifman M, Cheung KH, Miller PL, et al. X!!Tandem, an improved method for running X!tandem in parallel on collections of commodity computers. J Proteome Res. 2008;7(1):293–9.

19. Duncan DT, Craig R, Link AJ. Parallel tandem: a program for parallel processing of tandem mass spectra using PVM or MPI and X!Tandem. J Proteome Res. 2005;4(5):1842–7.
20. Kim S, Pevzner PA. MS-GF+ makes progress towards a universal database search tool for proteomics. Nat Commun. 2014;5:5277.
21. Dorfer V, Pichler P, Stranzl T, Stadlmann J, Taus T, Winkler S, et al. MS Amanda, a universal identification algorithm optimized for high accuracy tandem mass spectra. J Proteome Res. 2014;13(8):3679–84.
22. Tabb DL, Fernando CG, Chambers MC. MyriMatch: highly accurate tandem mass spectral peptide identification by multivariate hypergeometric analysis. J Proteome Res. 2007;6(2):654–61.
23. Eng JK, Jahan TA, Hoopmann MR. Comet: an open-source MS/MS sequence database search tool. Proteomics. 2013;13(1):22–4.
24. Diament BJ, Noble WS. Faster SEQUEST searching for peptide identification from tandem mass spectra. J Proteome Res. 2011;10(9):3871–9.
25. Cox J, Neuhauser N, Michalski A, Scheltema RA, Olsen JV, Mann M. Andromeda: a peptide search engine integrated into the MaxQuant environment. J Proteome Res. 2011;10(4):1794–805.
26. Geer LY, Markey SP, Kowalak JA, Wagner L, Xu M, Maynard DM, et al. Open mass spectrometry search algorithm. J Proteome Res. 2004;3(5):958–64.
27. Tharakan R, Martens L, Van Eyk JE, Graham DR. OMSSAGUI: an open-source user interface component to configure and run the OMSSA search engine. Proteomics. 2008;8(12):2376–8.
28. Nahnsen S, Bertsch A, Rahnenfuhrer J, Nordheim A, Kohlbacher O. Probabilistic consensus scoring improves tandem mass spectrometry peptide identification. J Proteome Res. 2011;10(8):3332–43.
29. Balgley BM, Laudeman T, Yang L, Song T, Lee CS. Comparative evaluation of tandem MS search algorithms using a target-decoy search strategy. Mol Cell Proteomics. 2007;6(9):1599–608.
30. Vaudel M, Burkhart JM, Zahedi RP, Oveland E, Berven FS, Sickmann A, et al. PeptideShaker enables reanalysis of MS-derived proteomics data sets. Nat Biotechnol. 2015;33(1):22–4.
31. Koenig T, Menze BH, Kirchner M, Monigatti F, Parker KC, Patterson T, et al. Robust prediction of the MASCOT score for an improved quality assessment in mass spectrometric proteomics. J Proteome Res. 2008;7(9):3708–17.
32. Yates 3rd JR, Eng JK, McCormack AL, Schieltz D. Method to correlate tandem mass spectra of modified peptides to amino acid sequences in the protein database. Anal Chem. 1995;67(8):1426–36.
33. Colinge J, Masselot A, Giron M, Dessingy T, Magnin J. OLAV: towards high-throughput tandem mass spectrometry data identification. Proteomics. 2003;3(8):1454–63.
34. Nesvizhskii AI, Keller A, Kolker E, Aebersold R. A statistical model for identifying proteins by tandem mass spectrometry. Anal Chem. 2003;75(17):4646–58.
35. Han X, He L, Xin L, Shan B, Ma B. PeaksPTM: mass spectrometry-based identification of peptides with unspecified modifications. J Proteome Res. 2011;10(7):2930–6.
36. Huang X, Huang L, Peng H, Guru A, Xue W, Hong SY, et al. ISPTM: an iterative search algorithm for systematic identification of post-translational modifications from complex proteome mixtures. J Proteome Res. 2013;12(9):3831–42.
37. Nesvizhskii AI, Roos FF, Grossmann J, Vogelzang M, Eddes JS, Gruissem W, et al. Dynamic spectrum quality assessment and iterative computational analysis of shotgun proteomic data: toward more efficient identification of post-translational modifications, sequence polymorphisms, and novel peptides. Mol Cell Proteomics. 2006;5(4):652–70.
38. Bern M, Kil YJ, Becker C. Byonic: advanced peptide and protein identification software. Curr Protoc Bioinformatics 2012;Chapter 13:Unit 13.20
39. Taus T, Kocher T, Pichler P, Paschke C, Schmidt A, Henrich C, et al. Universal and confident phosphorylation site localization using phosphoRS. J Proteome Res. 2011;10(12):5354–62.

40. Mitchell A, Chang HY, Daugherty L, Fraser M, Hunter S, Lopez R, et al. The InterPro protein families database: the classification resource after 15 years. Nucleic Acids Res. 2015;43(Database issue):D213–21.
41. Druce M, Hulo C, Masson P, Sommer P, Xenarios I, Le Mercier P, et al. Improving HIV proteome annotation: new features of BioAfrica HIV proteomics resource. Database. 2016;2016:baw045.
42. Hrecka K, Hao C, Gierszewska M, Swanson SK, Kesik-Brodacka M, Srivastava S, et al. Vpx relieves inhibition of HIV-1 infection of macrophages mediated by the SAMHD1 protein. Nature. 2011;474(7353):658–61.
43. Yang W, Jackson B, Zhang H. Identification of glycoproteins associated with HIV latently infected cells using quantitative glycoproteomics. Proteomics. 2016;16(13):1872–80.

Printed in the United States
By Bookmasters